Studies in
PLANT PHYSIOLOGY SERIES, **No. 1**

PLANT SENESCENCE

Physiological, Biochemical and Molecular Aspects

Studies in Plant Physiology Series No. 1

PLANT SENESCENCE:

Physiological, Biochemical and Molecular Aspects

R. Ranjan
S. P. Bohra
M. Jeet Asija

Agrobios (India)

Dr. Rajiv Ranjan
Department of Botany
T. P. Verma College
Narkatia Gunj 845 455 (Bihar)

Dr. S. P. Bohra
Laboratory of Stress Physiology
Department of Botany
J. N. V. University, Jodhpur (India)

Dr. Miss Mahendera Jeet Asija

Agrobios (India)

9 788177 540796
Rs. 1500.00 / US$ .00
Published by:
AGROBIOS (INDIA)

ISBN [illegible]-3

Price: Rs. [illegible] 40.00

Published by: Dr. Updesh Purohit for Agrobios (India), Jodhpur
Lasertypeset at: Commerce House, Jodhpur
Cover Design by: Reena
Printed at: Anmol Prints, Jodhpur

PREFACE

Senescence is a suicidal impulse. It is a programmed cell death occurs universally. Senescence is a genetically programmed sequence of biochemical and physiological changes. The initiation of senescence is subject to regulation by various environmental and autonomous (internal) factors.

Plants have evolved mechanisms by which leaf senescence can be induced by many stresses to reallocate nutrients to reproductive organs and to eliminate water consumption by older, less productive leaves. This regulation of leaf senescence has an obvious adaptive value, allows in the plant to complete its life cycle even under stressful condition. Senescence is a correlative event and is the cumulative effect of the formation of flowers, fruits, seeds, apices and roots.

Senescence is a mechanism, which works in favour of maintaining an ecological balance both for the internal milieu and the external environment of a species.

The present book covers various aspects of plant senescence including ecological, physiological, biochemical and molecular aspects. We have tried to cover the subject in light of current researches.

We are thankful to Dr. S. S. Purohit and Dr. U. Kumar for help in various ways.

Authors

CONTENTS

PHYSIOLOGICAL AND MOLECULAR ASPECTS REGULATING SENESCENCE

ENVIRONMENTAL REGULATION OF SENESCENCE

Chapter 1

SENESCENCE: A SUICIDAL IMPULSE

One of the most puzzling events in a living system is its death. Plants and their parts develop continuously from germination to death. The latter part of the developmental process that leads from maturity to the ultimate complete loss of organization and function is termed senescence. In living organisms there occur **programmed cell death** (PCD). It refers to a process by which cell promotes their own death through the activation of self-destruction system.

PCD is a physiological cell death process involved in the selective elimination of unwanted cells. In animals these unwanted cells include those that have served temporary functions, cells that are overproduced and cells that are unwanted or present in inappropriate positions. PCD in animals depends on the induction and action of specific genes that bring about the controlled disassembly of a cell. It involves the condensation, shrinkage and fragmentation of the cytoplasm and the fragmentation of the nuclear DNA into 50 kb and in some cases 0.14 kb pieces that represent oligonucleosomes. These fragments can be detected by terminal deoxy-nucleotidyl transferase – mediated dUTP nick end labelling (TUNEL) of DNA 3′ –OH groups in sections of cells. Many of the proteins responsible for PCD have now been identified. Proteases of the interleukin - 1β - converting enzymes (ICE) family bring out cell condensation and shrinkage and endonucleases causes the fragmentation of nuclear DNA. The basic leucine Zipper transcription factor, ces-2 (cell death specification-2) and oncoproteins in the Bcl-2 family regulate the onset of PCD in many animal cells. Growth factor deprivation and reactive oxygen species can trigger PCD. Ca^{2+} also participates in PCD.

It is now generally accepted that many plant developmental processes and stress responses are achieved through the operation of PCD. These include senescence, xylogenesis, embroygenesis, sex determination in monoecious plants such as maize, abscission zone formation and the hypersensitive reaction to pathogen infection. In plans selective cell death is necessary for growth and survival and can occur on local or large scale. Morphological observations of dead and dying cells show that the cytoplasm and nuclei condense and shrink. TUNEL experiments reveal oligonucleosome sized DNA fragments that are characteristic of PCD. It appears that the mechanism controlling PCD in plants and animals may be similar. However, that is no molecular evidence for this.

During the process of cell death, various proteins that are normally sequestered in the intermembrane space of mitochondria including Cytochrome C, apoptosis-inducing factor and certain procapspase proteins are released into cytosol. Depending upon the stimulus, this event can lead to necrosis through irreversible mitochondrial damage and energetic catastrophe C-dependent Caspase activation. In the cytosol, cytochrome C activates the caspase, a family of killer protease through formation of a complex with Apaf-1 (apoptotic-protease activating factor-1), procaspare-g and ATP or dATP (Martinou *et al.*, 2000). Release of cytochrome C from mitochondria during apoptosis is controlled by proteins of the Bcl-2 family: those that inhibit cell death and prevent the release of cytochrome C. Little is known about the mechanism by which cytochrome C escapes from mitochondria during cell death. It is reported that the release of cytochrome C is a rapid, complete and kinetically invariant event.

Animal and plant senescence differ in a variety of ways although both are apparently the manifestations of the same basic phenomenon. In animals senescence is largely due to the cumulative deleterious effects of the environment upon the organism. In contrast, in plants it is largely hereditary. This is because the animals, which due to their heterotrophic mode of nutrition are rather a delicate system, depend upon other sources for food as against plant which being autotrophs are rather hardy and relatively less liable to the harmful inflictions of the environment. In most instances, plants senesce not because of adverse climatic conditions but due to and after the ripening of fruits, since they start senescing even at a time when climatic conditions are still favourable for their normal growth.

Senescence is a genetically programmed sequence of biochemical and physiological changes. Its initiation is subject to regulation by many environmental and autonomous (internal) factors. The environmental include stresses such as extremes of temperature, drought, ozone, nutrient deficiency, pathogen infection, wounding and shading where as the antonomous factors include age, reproductive development and phytohormone levels (Fig. 1.1).

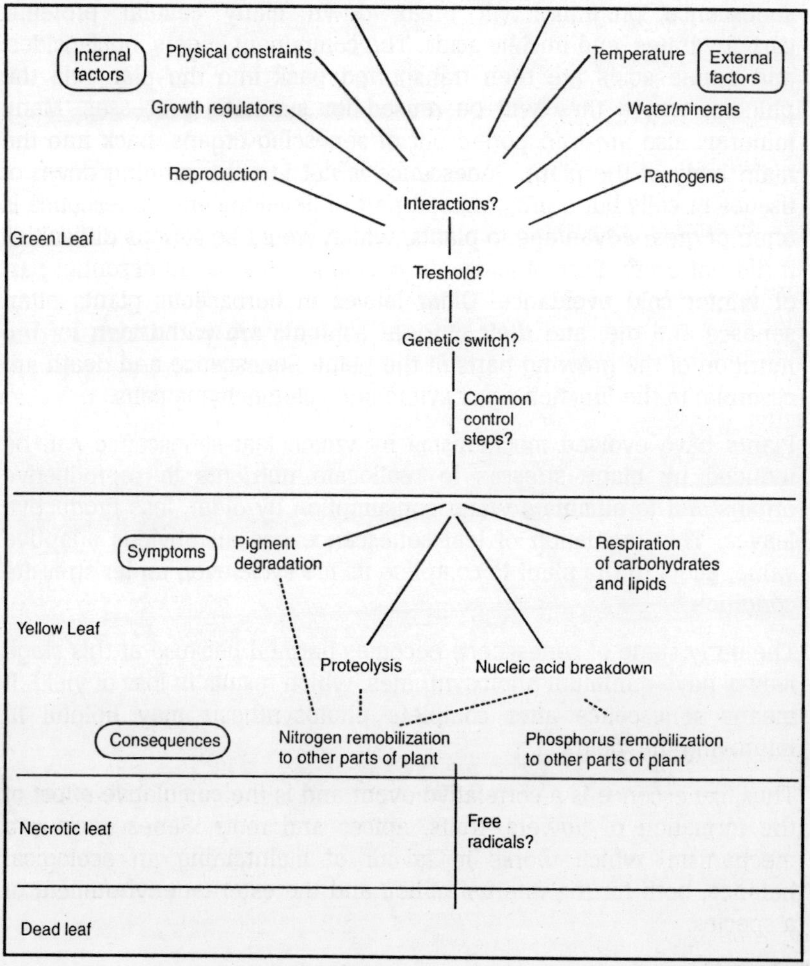

Fig. 1.1: Factors implicated in the initiation of leaf senescence (After Smart, 1994).

Senescence is an essential aspect of plant development as after the senescence leaves die and withdraw their photosynthate to reproductive part. As new leaves are initiated from the shoot apical meristem, older leaves often are shaded and lose the ability to function efficiently in photosynthesis. Since the plant invested valuable resources in leaf formation, senescence recovers a portion of those resources. Hydrolytic enzymes, produced as part of the senescence program, will break down many cellular proteins, carbohydrates, and nucleic acids. The component sugars, nucleosides, and amino acids are then transported back into the plant via the phloem, where they will be reused for synthetic processes. Many minerals also are transported out of senescing organs, back into the main body of the plant. Senescence is not just the running down of tissues or cells but a programmed part of development. Senescence is often of great advantage to plants, which would be serious difficulties if did not occur. Loss of leaves in deciduous trees is an essential part of winter cold avoidance. Older leaves in herbaceous plants often senesce and die, and their nutrient contents are withdrawn for the nutrition of the growing parts of the plant. Senescence and death are essential to the functioning of xylem and sclerenchyma cells.

Plants have evolved mechanisms by which leaf senescence can be induced by many stresses to reallocate nutrients to reproductive organs and to eliminate water consumption by older, less productive leaves. This regulation of leaf senescence has an obvious adaptive value, allow in the plant to complete its life cycle even under stressful condition.

The early stage of senescence becomes harmful because at this stage leaves have minimum photosynthates, which results in loss of yield. It means senescence after complete photosynthesis may helpful in improving the yield.

Thus, senescence is a correlative event and is the cumulative effect of the formation of flowers, fruits, apices and roots. Senescence is a mechanism, which works in favour of maintaining an ecological balance, both for the internal milieu and the external environment of a species.

IS SENESCENCE IS A UNIVERSAL PHENOMENON?

Processes that are similar to programmed cell death have been described in prokaryotes and lower eukaryotes. Some strains of *E. coli* harbor plasmid encoded genes that trigger proteolysis and cell death after infection by bacteriophage (Ameisen, 1996) and activation of the AMP pathway in non-differentiating cells of Dictyostelium discoides induces cell death process closely resembling programmed cell death. Culture of several kinds of unicells at low density also induces a process resembling programmed cell death (Welburn et al. 1996). Programmed cell death is also thought to occur in the green alga *Volvox* and the fern *Marsilea*. Degreening in *Chlorella* is accompanied by the loss of proteins and is associated with ultrastructural changes of the chloroplasts such as the loss of thylakoids and production of granular bodies. These changes resemble the changes occurring during the transition of chloroplasts into gerontoplasts in leaf senescence.

Its universality has been questioned primarily because *Chlorella, Chlamydomonas* and other unicellular organisms including prokaryotes like bacteria can be kept indefinitely in media or chemostatic cultures without any exhibition of the symptoms of senescence. If so, then why do multicellular forms, obviously derived from these so-called immortal unicellular forms, undergo senescence? One plausible explanation is that a multicellular body is integration or various cellular units assigned with particular function or in simple terms, the division of labour. In such an integrated and inter-associated complex of several systems even a slight damage to one part will show long-range effects. For example in a tracheophyte, if the ascent of sap is affected even to growth of the whole plant will be affected. Hence it is the accumulation of immobile by products of metabolism in tissues that often results in progressive decline in vigour, in addition to the genetical factors. The rate of metabolism can be directly related to senescence. The more rapidly a living system metabolizers, the earlier would be the arrival of senescent phase. The only exception to this is the phenomenon of hybrid vigour. Although the unicellular forms do not exhibit senescence in *sensu stricto*, organelle senescence in them cannot be denied. Another reason why unicellular forms do not show senescence is their constant cell renewal. It is a fact that meristematic cell do not age.

AGING *Vs.* SENESCENCE

According to Medawar (1957) senescence may be defined as the *deteriorative processes that are natural causes of death.* Aging, by contrast, refers to processes of occurring maturity with the passage of time. By these definitions, then aging includes a much wider span of physiological changes, some of which may lead to the weakening of the organism while others may be quite neutral with respect to capability of the biological organism to survive. Senescence, on the other hand, refers to those changes that provide for the endogenous regulation of death. As examples, the physiological changes in a plant, which cause its conversion from a seedling to juvenile plant, from a juvenile plant to a mature plant, or the gradual decline in vigor with increasing age, would be examples of aging. On the other hand, physiological changes involved in the autumnal colouring (and subsequent death) of leaves, or in the dieback of tulips and other spring bulbs in early summer, or in the yellowing and death of annual and biennial species after the completion of fruiting or after a brief period of drought would be examples of senescence.

Relationship between aging and senescence is also worth considering. Aging is mostly taken to be a continuous process from embryogenesis to death, whereas senescence always starts after reaching the reproductive maturity. However, some people still take aging and senescence as synonyms.

Longevity is a natural correlate of aging and senescence. The ranges of longevity over the plant kingdom are enormous, ranging from a few weeks (in the case of desert annuals) to perhaps 4000 years (in case of *Bristlecone* pines). Generally, it might be said that the shortest-lived plant species tend to have longevities limited by programmed senescence, whereas, the perennials tend to have much less precise limits of longevity and tend to die through a gradual attribution associated with aging.

VARIOUS LEVELS OF SENESCENCE

Senescence can be expressed at various levels.

Organism senescence

It is a kind of senescence terminalized by the death of the plant. The life span varies to a very great extent from about a few days to 4,500 years. Annuals and biennials have a very short life span. They lack elaborate modifications to escape detrimental effects of the forthcoming unfavorable season. Their flowering is determined by the photoperiodic stimulus and it finally results in the death of individual. Their growth curve is typically sigmoid, consisting of a lag phase with little or no growth, a log phase with steep increase in the growth rate, a period of zero growth or growth plateau, followed by a negative curve or the phase of senescence. Trees are unique in avoiding the detrimental effects of the season by shedding their leaves. The life span varies a great deal among trees. *Betula* starts senescing at the age of 50 years, Redwood (*Sequoia*) lives for 3000 years, and one specimen of Bristlecone Pine has been proved to be 4,500 years old. Trees show a staircase like growth curve: log phase represented by the spring season growth a growth plateau by the autumn growth. Trees in general show a continuity of life span for thousands of years, but it is led by a rapid death of cells followed by continuous cell replacement by meristems.

Clonal senescence

Clonal senescence is decline in vigour through successive generations in vegetatively propagated species. Its occurrence still rests in question. *Cryptomeria, Musa* and so many other fruit trees have been grown for centuries by stem cuttings, still no aging symptoms are reported. However, there is a reasonable evidence for aging in *Citrus* clones.

Organ senescence

Shedding of cotyledons from a seedling in epigeal germination, periodic shedding of leaves in deciduous trees, shedding of floral parts (calyx, corolla, stamens etc.) and fruit ripening come under this category.

Tissue senescence

It is an important part of tissue differentiation. Cork formation by suberization, sclerenchyma formation in general, and xylogenesis, all are outcome of this process.

Cell senescence

Cell is the basic unit of life and cells are involved in all kinds of senescence, but the senescence of a cell in particular is also met with. It may be a part of differentiation, for example sclereid differentiation, cork cell formation or may be due to some irrepairable defect arising in a cell. In the latter case the cell simply undergoes autophagic digestion without causing any harmful effects to neighbouring cell. Cells with longest life span so far reported are the pith cells of *Carnegia gigntea*, which live for about 100 years. In trees, the pith and ray parenchyma cells are the oldest living cells.

Organelle senescence

All living cells constantly show organelle senescence and organelle renewal, even in the so-called immortal unicellular forms. As an organelle falls in, due to some irrepairable changes or defects, it undergoes senescence and is finally digested by autophagy. One of the primary and easily seen phenotypic symptoms of senescence is yellowing of leaves. Cotyledons and fruits also show yellowing with the arrival of senescent phase. Another common symptoms are the loss of rigidity of stems, coupled with yellowing in dicotyledonous herbs and some monocots. Rigidity of leaves and fruits are also considerably decreased due to hydrolysis of organizational framework of the plant.

MONOCARPY *Vs.* POLYCARPY

In considering the senescence of the whole plants, most species can be classified into two groups.

- **Polycarpic plants.** Plants that normally have more than one reproductive phase before death.
- **Monocarpic plants.** Plants that have one reproductive phase followed by death.

According to Woolhouse (1982) monocarpy has polyphylatically evolved from polycarpy through natural selection and hybridization. Indeed, crop plants (mostly monocarpic) are usually developed from their monocarpic progenitors through artificial selection in breeding programme where in most cases achievement of increase in the ratio of reproductive: vegetative portion at harvest is an obvious goal.

However, the above ratio is found to be lower in a wider range of polycarpic plants in a particular season (personal observation). Polycarpic plants are subjected to annual organ as well as cellular formation of secondary growth leading to the formation of dead cork cells, tracheae etc.). Senescence is lacking in herbaceous monocarpic individuals. It may be speculated that efficient processes of recycling of nutrients along with periodic detoxification through abscission of leaves and other useless organs (which might act as a potential portal for infection) in the polycarpic plants provide means for greater survival compared to the monocarpic plants. Perhaps stem of the polycarpic deciduous plants serves as an efficient intermediary storage organ where nutrients are temporarily stored after recovery from the senescing leaves for further use by the newly developed foliar organs. In some cases monocarpic plants also possess intermediary storage organs (stem, fruit wall etc., Biswas and Mandal, 1987b) where nutrients from leaves are temporarily stored.

Monocarpy occurs in a wide range of plant taxa and evidently has evolved independently numerous times. Secondly, while many annuals and biennials are monocarpic, some perennials (e.g., bamboo century plant) are also monocarpic. Thirdly, monocarpy is very common in cultivated plants, especially field crops, where the abbreviated life span and complete desiccation prior to harvest and/or synchronous production have apparently been selected (Schwanitz, 1966). Monocarpic field crops in fact put on the world's most dramatic displays of senescence, for vast fields of individuals undergo senescence and die essentially simultaneously. It is convenient to call this degeneration following the reproductive phase, monocarpic senescence. In monocarpic (those that flower only once and then die) senescence (Noodén and Lindoo, 1978) the entire plant senesces at the time of reproductive development. Leaf senescence precedes the fall of dead leaves in deciduous trees. The maturation of xylem cells derived from vascular cambium may be considered as a senescence process, which is under hormonal control. However, it is difficult to state definitely what processes naturally terminates the life of an organ or a plant and what processes are simply reflection of the changes associated to the passage of time.

Plants do not grow in pieces. They age, as they grow and develop, in orderly fashion. In some plants it occurs as the gradual enrichment of deterioration process and in other plants there may be a fairly abrupt deterioration leading to death. The study of plant senescence is concerned with deteriorative changes which occur during plant development and which result in death of either entire plant or portion of the plant. All cells and tissues in the plant ultimately undergo senescence.

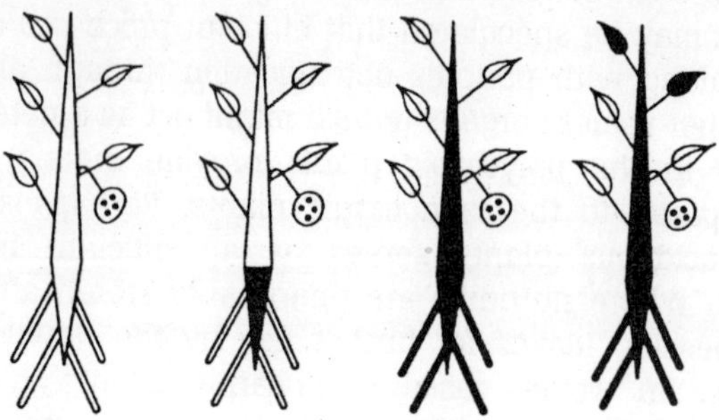

Fig. 1.2: Representation of various patterns of senescence in plants.

According to Leopold (1961) there are four morphological patterns of senescence in plants, viz., overall senescence-where the individual dies in an abrupt way (most of the monocarpic plants); top senescence-where senescence of the aerial parts of the plants occurs; deciduous senescence-where it includes the annual senescence of the leaves as in woody perennials; and progressive senescence-where it concerns the sequential senescence of the leaves in an acropetal order.

Senescence of the entire plant after a single reproductive cycle is called monocarpic senescence. Other types of senescence include the senescence of aerial shoots in herbaceous perennials, seasonal leaf senescence (as in deciduous trees), sequential leaf senescence (in which the leaves die when thcy reach a certain age), senescence of dry fruits, senescence (ripening) of fleshy fruits, senescence of storage cotyledons and floral organs, and senescence of specialized cell types (e.g., trichomes, tracheids, and vessel elements).

Senescence can occur at the level of the whole plant, as in monocarpic senescence; at the organ level, as in leaf senescence; and at the cellular level, as in tracheary element differentiation. The process whereby individual cells activate an intrinsic senescence program is called programmed cell death (PCD).

Senescence occurs in a variety of organs and in response to many different cues. Many annual plants, including major crop plants such as wheat, maize, and soybeans, abruptly yellow and die following fruit production, even under optimal growing conditions. Senescence of the entire plant after a single reproductive cycle is called **monocarpic senescence**. Other types of senescence include the senescence of aerial shoots in herbaceous perennials, seasonal leaf senescence (as in deciduous trees), sequential leaf senescence (in which the leaves die when they reach a certain age), senescence of dry fruits, senescence (ripening) of fleshy fruits, senescence of storage cotyledons and floral organs, and senescence of specialized cell types (e.g., trichomes, tracheids, and vessel elements).

The entire plant development can be viewed as one series of events. The successive stages of development, starting with juvenility, vigorous vegetative growth at its peak, maturity of plant when mobilization of nutrients out of the vegetative parts into the flowers and fruits occurs and all this followed by senescence, are attended by numerous advantages. During senescence, this process is imposed in regular patterns due to widespread occurrence of cellular and organ senescence, which slowly but steadily spreads through the entire plant in a natural fashion.

Chapter 2

PLANT GROWTH AND SENESCENCE

During the course of their lives, all multicellular organisms and their organs and tissues reach a peak in terms of their physiological function, and then they decline until they die. According to Molisch (1938), the death of the whole plant commonly occurs (annual and biennial species) upon completion of fruiting. The reproductive activity is lethal to the plant and removal of the flower and fruit of several species of plants prevents senescence.

Senescence is almost certainly a complex of many sequential and parallel steps. Cessation of vegetative growth, particularly stem elongation or senescence of apical meristem, and the transition from juvenile to adult phase though interesting phenomena, may not be measures of whole plant senescence. Among the different types of senescence (organelle, cellular, organ and whole plant), the whole plant senescence in monocarpic annuals seems to be unique in that further development of the plant ceases in an abrupt but coordinated way, seemingly for the formation of reproductive propagules, while in cellular along with their organelles (during the formation of sieve tube, tracheae, lysigenous cavity, etc.) and organ senescence (abscission) further developmental processes are tagged, whole plant senescence terminates the life process providing means for adaptation (e.g. in ephemerals) and evolution (e.g. in annuals).

EFFECT OF SENESCENCE ON VEGETATIVE GROWTH

Whole plant senescence is accompanied or preceded by a decrease or cessation of both shoots and root growth. The reduction in vegetative growth that occurs during the reproductive phase has been recognized for a long time in both monocarpic and polycarpic plants, and is generally attributed to a shift in distribution of growth limiting nutrients within the plant (Leonard, 1962; Loomis, 1953; Pearsall, 1923; Wangermann, 1965). Thus, it is of importance to determine when the reduction in root or shoot growth occurs relative to flowering, fruit development, and senescence. Where there shoot apex is converted to an inflorescence (e.g., corn and cocklebur), vegetative node production ceases coincident with the appearance of flower buds, though stem elongation may continue. In other species (e.g., pea and cotton) where the apical meristem of the main shoot does not become and inflorescence, it may stop growing and senescence during their reproductive phase. In soybean, (Borst and Thatcher, 1931; Lindoo and Noodén, 1977; Noodén, 1977) cotton (Crowther, 1935) and pea (Lockhart, 1961), stem elongation decreased but continues after flowering has started and stops while the fruit are growing

Flowering and vegetative growth are not always antagonistic, for under long days (16 hr), the shoot apex of morning glory, *Ipomoea caerulea*, continues vegetative growth and flowers are produced from lateral buds, but short days cause the terminal bud to become a flower (Ashby, 1950). Similarly, Marx (1958) has found a variety of pea (G2) that continuous to grow and produce fruit under short days, whereas these plants terminate growth and die under long days. It is important to know if the rate of fruit development and nutrient demand (e.g., dry weight or nitrogen accumulation in seeds) is the same in senescing and non-senescing plants. In soybean root growth has been reported to terminate at about the same time as shoot growth (Mitchell *et al.,* 1971) or a bit sooner. (Borst, 1931) Cotton root growth also declines during the reproductive phase, apparently ahead of shoot growth. The roots of radish cease growth at flowering. The roots of common perilla (*Perilla frutescens*), continues to grow after node production ceased, but start to slow about the time when the first flowers open (Beever and Woolhouse, 1975). In rice and wheat (Asana and Singh, 1967; Brenehley and Jackson, 1921) the roots may stop growing later, during heading, and degenerate even lateral during grain development. Likewise, corn roots grow little after seed growth has started (Foth, 1962). Plentiful supply of water and nutrients may however, sustain wheat roots. (Carpenter *et al.,* 1952) Pea nodules reach a maximum mass at flowering, (Pate, 1958) and substantial losses may occur at fruiting. (Pate, 1958; Salter and Drew, 1965) Soybean nodule also found to senesce during fruit development. (Hardy *et al.,* 1973) Thus, at least in some cases, root growth may be affected early in reproductive development, apparently too early to due to a nutrient deficiency resulting from massive diversion of the nutrient supplies to developing fruit.

The prime candidates for control centers for growth cessation are the reproductive structures. Indeed, the nutrient diversion explanation of the cessation of growth carries with it the prediction that removal of the reproductive sinks should restore vegetative growth, but this is only partially true.

In the most clear-cut cases, *Viena wallflower* (Reichert, 1938) and *Mignonette,* (Molisch, 1938) simple defloration prevents growth cessation, and the plants become very large. At the other extreme, it has been reported that defloration does not restore vegetative growth or even prevent senescence in cocklebur (Krizek, 1966) In cotton (Dale, 1959; Eaton, 1931; Eaton, 1952; Mason, 1922) okra (Harvey, 1931) tomato (Murneek, 1926) corn, (Loomis, 1935) tobacco, (Petrie, 1939) pea, (Ecklund, 1968; Ecklund, 1969; Lockhart, 1961; Malik, 1973; Malik, 1976; Pate, 1958) common perilla (Beever and Woolhouse, 1975) pepper, (Hall, 1977) and others (including some grasses that are used in lawns). Molisch, (1938) defloration or defruiting may prolong or promote vegetative growth. Defruiting (Lindoo and Noodén, 1978) or even deflowering only slightly promotes or prolongs stem elongation in soybean, although the treated plants have thicker, greener leaves and a more robust appearance (even when compared with untreated plants at their peak) On the other hand, flower removal may temporarily depress elongation of cucumber vine (Davis and McCree, 1977). Cessation of growth, including stem growth, does seem to relate to monocarpic senescence, and in some plants, this may be a convenient measure of senescence. Cessation of apex growth, however, does not cause the death of most monocarpic plants, but may merely accompany monocarpic senescence.

EFFECT OF SENESCENCE ON REPRODUCTIVE GROWTH

Many plants produce far more flowers than fruit and initiate more fruit than they carry to maturation. Considerable evidence from a wide range of monocarpic and polycarpic plants indicates that the reproductive structures themselves exert an inhibitory effect on the development of other reproductive structures and may even induce their abortion, deflowering or depodding can offset these effects.

Normally, this inhibition of excess flower and fruit development does not cause whole plant senescence. Nonetheless, it seems important to determine whether or not the flowers or fruit produce these effects via the same mechanism by which they induce senescence or inhibit vegetative growth. In the case of soybeans, at least, there seem to be some differences between these inhibitory effects. For example, continuously deflowered plants produce several times more flowers than control plants (260 and 100 flowers per plant respectively), and they are produced over a much longer period in deflowered plants, whereas pod removal has much less promotive effect on flower initiation (Noodén, 1977). Thus, soybean flower initiation is controlled mainly by the flowers, or very young pods, in contrast to senescence, which is induced by the pods during their later phase of development (Noodén, 1977). Since pod initiation and abortion do not seem to be altered by the auxin-cytokinin treatment, which prevents monocarpic senescence in soybeans, (Noodén, 1979) it appears that the senescence signal in soybeans may not repress pod initiation or promote abortion. In the soybean, the developing reproductive structures do not seem to control the termination of stem elongation (Lindoo, 1976; Noodén, 1977) but the pods do induce monocarpic senescence (Lindoo, 1977) Thus a variety of different correlative signals seem to be involved, at least in soybeans.

SENESCENCE RELATED TO STRESS

A new fad is abroad in the biological world- stress physiology. There are those who would put disease-induced senescence among the stress induced phenomena. It involves mechanical damage, chemical damage, and the metabolic load associated with the presence of pathogen, etc. None of these is a well defined phenomenon at the biochemical level.

However, these factors do play a role in most diseased tissue and they are important from the point of view of the plant pathologist and physiologist because they lead to striking changes in metabolism; they can easily obscure the more subtle, specific changes and are often related to symptom develop. They are responsible for the fact that some symptoms, e.g., senescence, are widespread in disease plants.

MODELS PROPOSED FOR MECHANISM OF SENESCENCE

It is difficult to propose any single mechanism of senescence in chloroplasts for two main reasons. First, it is difficult to define a senescence mechanism because of the contradictions in the accumulated data resulting from system specificity and a lack of uniformity in experimental conditions. Second, our knowledge of chloroplast senescence is still inadequate and the available data are not sufficient to develop concepts, generalize them, and propose a definite explanation of organelle degradation. Three possible mechanisms are discussed.

The loss of the efficiency of the Calvin cycle as the first major step during chloroplast senescence. The signals for the loss of Rubisco, the crucial enzyme of the Calvin cycle, comes from the to oxygen to generate oxygen free radicals such as superoxide that cause peroxidation of thylakoid lipids including lipids on the oxygen-evolving complex.

Damage to the oxygen-evolving system results in the generation of singlet oxygen and other free radicals. Free radical-induced lipid peroxidation may activate membrane-specific of membrane lipids. Alternatively, a loss of lipids may expose intrinsic membrane proteins and make them accessible to already existing active proteases. Lipid peroxidation may inactivate membrane-bound enzymes and electron-transport carriers by oxidation and cross-linking of proteins. This may ultimately lead to complete breakdown of the chloroplast. The difficulty in accepting the Calvin cycle as the first major step of chloroplast degradation is the lack of data on the activity of Rubisco *in vivo* during organelle senescence.

Most of the results regarding the rapid loss of the enzyme activity is measured *in vitro*, which may not simulate the conditions occurring in leaves. In addition, there are data available from some plant systems that suggest that the loss of chlorophyll and the electron-transport system occurs before any significant loss of Rubisco activity which would mean that a loss in the efficiency of the Calvin cycle may not be a limiting factor for chloroplast senescence. It is possible that a senescence-specific loss of thylakoid lipids triggered by a lipase is the first step of chloroplast degradation. The oxygen-evolving system has a very delicate structural organization and is highly susceptible to even mild stress. The role of thylakoid lipids in oxygen evolution has been established (Akabori *et al.,* 1984). The loss of lipids may cause the release of extrinsic proteins and/or Mn^{2+} from the oxygen-evolving complex. The release of extrinsic proteins under stress of Mn^{2+} during senescence have been reported. This obviously leads to a decline in electron-transport efficiency and a change in the light-induced concentration of Mg^{2+} and pH of the stroma, which in turn affect the efficiency of carbon dioxide fixation in the Calvin cycle.

There is a loss of some of the components of the electron-transport chain linking the oxygen-evolving complex to NADP reduction. The loss of the efficiency of the electron-transport system is attributed to the quantitative loss of quinones, reaction centers, plastocynin, Fd-NADP reductase, or the coupling factor for the ATP synthesis. This may lead to loss of the efficiency of the water-splitting system or Calvin cycle resulting in the breakdown of the organelle.

Of the three above proposed schemes, the first one seems to be more logical than the other two. The rapid and early loss of Rubisco has physiological significance. Rubisco protein constitutes about 50% of the total protein of green leaves, the major source of nitrogen for other growing parts of the plant system. During chloroplast senescence, the degradation products of this protein move to the growing parts. However, transport of these products needs a continuous supply of ATP. It is, therefore, reasonable to assume relative stability of the electron-transport system over the Calvin cycle. When the economy of the system is considered, the second target of the degenerative process obviously should be the oxygen-evolving complex, which should degrade earlier than the intrinsic proteins of the electron-transport chain. As long as the oxygen-evolving system remains intact, it logically requires a totally efficient electron-transport chain to link NADP reduction and ATP synthesis. Therefore, in the sequence, the loss of intrinsic electron transport components should follow the senescence-induced damage to the water-splitting system.

Chapter 3

SENESCENCE INDUCED STRUCTURAL AND ULTRASTRUCTURAL MODIFICATIONS

Senescence is the last phase of development of a whole organism, organ, cell, or organelle. It is basically a degenerative process that leads to the death of a living system. Although its precise mechanism is not known, recent experiments suggest that the process is genetically programmed (Woolhouse, 1984). Senescence, although deteriorative in nature, is a requirement to facilitate turnover processes and is an essential process for growing parts of living organisms. Senescence of green leaves in plants has been extensively investigated in the last two decades (Thomas and Stoddart, 1980; Woolhouse and Jenkins, 1983; Biswal and Biswal, 1984). Different tissues and cells of the leaves have their own pattern and timing of senescence. The initiation of senescence in mesophyll cells may not necessarily be synchronized with the process in vascular or epidermal tissues. Even different organelles of a single leaf cell, namely chloroplasts, mitochondria, endoplasmic reticulum, ribosomes, and the nucleus do not show synchrony in the induction and progress of senescence (Biswal *et al.,* 1983b). Different organelles of a leaf cell degrade sequentially; the sequence may change depending on many factors (Woolhouse, 1984a). Chloroplasts are the first organelles to show symptoms of disorganization when all other organelles are normal (Dodge, 1970). This is followed by a change in the structure of the endoplasmic reticulum and loss of ribosomes. The next organelle that is affected is the mitochondrion, although at this time the nucleus, plasma membrane, and plastid envelope are shown to be intact. Mitochondria show symptoms of degradation that are reflected in the alteration and loss of the cristae and the reduction in the size of the organelle. Nuclear contents show shrinkage and the membrane becomes structurally irregular (Butler and Simon, 1971). Finally, senescence results in the rupture of the plastid envelope followed by the breakdown of the plasma membrane (Woolhouse, 1984a).

SENESCENCE INDUCED STRUCTURAL MODIFICATIONS

The deterioration processes in the leaf that leads to senescence begin as soon as the leaf attains its full size. Several types of structural changes are associated with leaf senescence. Before discussing the changes occurring due to senescence, we should deal with chloroplast structure.

Structure of chloroplast

The structure and organization of chloroplasts are described in detail by Kaplan and Arntzen (1982). The lamellar structure of chloroplast varies between photosynthetic organisms. The structure is relatively simple in less evolved plants such as blue-green algae and photosynthetic bacteria and more complex in higher angiosperms. Electron microscopic studies reveal the following basic features of a typical higher plant chloroplast (Fig. 3.1). The organelle is lens shaped and the lamellar system is encircled by a double membrane envelope. The envelope acts as a selective barrier to the transport of various metabolites into or out of chloroplasts. The lamellar system of the organelle is basically composed of lipoprotein disks that are stacked at regular intervals to form structures known as grana. The unstacked loose lamellae, which connect stacked grana zones, are intergranal or stroma lamellae (Fig. 3.1). The entire lamellar system is suspended in a granular stroma that is mainly composed of plastid genetic materials and ribosomes responsible for chloroplast replication and self-regulation. In addition, starch grains also constitute a part of the granular structure. The proteins present in the stroma are mostly the enzymes of the Calvin cycle. The other components of the stroma are plastoglobuli, which in electron micrographs appear as darkly stained structures consisting of lipids (Tevini and Steinmuller, 1985).

The grana and stroma lamellae are the structural units for primary photochemical reactions and contain the photosynthetic pigments in addition to lipids and proteins. The lipids of the thylakoid membrane consist of a complex mixture. The major lipids of the thylakoid membrane are galactolipids, phospholipids, and sulpholipids (Siegenthaler, 1982; Chapman *et al.*, 1985).

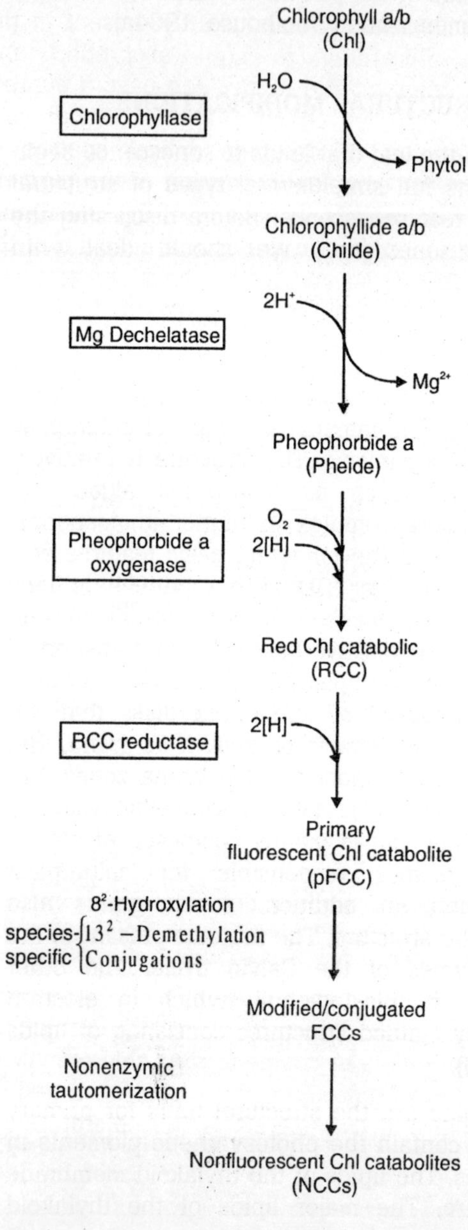

Fig. 3.1: The pathway of Chl breakdown in higher plants. Enzymes identified and characterized to date are boxed (after Hortensteiner, 1999).

Chlorophylls, carotenoids, and phycobilins are the major photosynthetic pigments and are involved in the process of light absorption by the organelle. Chlorophylls a and b, which are chemically different from each other, are abundantly present in the higher plant chloroplasts. According to light absorption and fluorescence characteristics, chlorophyll a *in vivo* can have a number of optical species as well as different chemical species (Rebeiz and Lascelles, 1982). The formation of different optical species of chlorophyll a is attributed to the complex interaction of the pigment with the membrane environment. A change in membrane environment could lead to an alteration in the optical behavior of the pigments. Therefore, light absorption and fluorescence spectroscopy are used as sensitive tools to probe the microenvironment of pigments in the membrane (Govindjee and Whitmarsh, 1982). Carotenoids and phycobilins found in lower plants are accessory pigments that absorb light energy and ultimately transfer it to chlorophyll for photochemical reactions. In addition, carotenoids protect chlorophyll molecules from photo-destruction. Chlorophyll and carotenoids are bound to thylakoid proteins to form different types of pigment-protein complexes (Kaplan and Arntzen, 1982; Ortiz *et al.*, 1985).

Proteins associated with the thylakoid membrane are coupling factor proteins, light-harvesting chlorophyll-protein (LHCP) complex, proteins of the electron transport carrier, proteins associated with the reaction center and the polypeptides of the oxygen-evolving complex. The proteins of the coupling factor and some of the polypeptides of the water-splitting system are extrinsic, but most of the thylakoid proteins are intrinsic in nature.

LOSS OF PIGMENTS, LIPIDS, AND MACROMOLECULES

Leaf senescence causes a loss of all types of photosynthetic pigments. However, depending on the plant system and environmental conditions, differential rates of loss of chlorophyll *a*, chlorophyll *b*, and carotenoids have been demonstrated (Sestak, 1977; Woolhouse and Jenkins, 1983; Thomas, 1983; Grover *et al.*, 1986). Carotenoids were shown to be relatively stable in comparison to chlorophylls. In contrast, equal rates of degradation have been demonstrated for chlorophyll a, chlorophyll b, and carotenoids of the chloroplasts of fern leaves (Biswal *et al.*, 1983).

Leaf senescence is characterized by a massive loss of green pigments (Biswal *et al.,* 1984; Smart, 1994), a process referred to as degreening. The degreening process of mature leaves that occurs either in nature or induced by darkness in laboratory conditions is accompanied by the disorganization of various cellular and subcellular structures (Biswal *et al.,* 1988, Smart, 1994).

Chlorophyll breakdown in senescent leaves

The biosynthetic pathway of Chl probably the most pigment in the biosphere, has been studied in considerable detail and is mentioned in plant physiology textbooks. Oddly enough, however, the breakdown of Chl is barely mentioned. Yet, as pointed out by Hendry et al. (1987), an estimated one billion tons of Chl are broken down every year, seemingly without leaving a trace. The only enzyme that is thought to be chlorophyllase, discovered by Arthur Stoll in 1912, which cleaves Chl into phytol and Chlide the Mg-porphyrin moiety of Chl. For most of the subsequent 80 years, chlorophyllase has been the only known part of the catabolic system. Until quite recently all the subsequent steps of Chlide breakdown remained undiscovered because no one was able to identify the break-down products.

Most of the biochemical steps of the catabolic pathway of Chlorophyll are now known. The catabolism of chlorophyll has been studied in senescing leaves and ripening fruits of a large number of plant species. The pathway of chlorophyll degradation as depicted in Fig. 3.2 summarizes current knowledge. The catabolic machinery comprises at least six reactions, whereby the first four are responsible for the conversion of chlorophyll to a primary colourless catabolite (pFCC). Chlorophyll is first dephytylated to chlide by the action of chlorophyllase and subsequently the central Mg atom is removed by Mg dechelatase. The product of the reaction, Pheide, is the last coloured (green) intermediary catabolite of the pathway. Its porphyrin macrocycle is oxygenolytically opened in a two-step reaction with a red catabolite (RCC) as an intermediate, the product, pFCC, is a fluorescent linear tetrapyrrole. This key step in chlorophyll breakdown is catalyzed by the joint action of Pheide a oxygenase (PaO) and RCC reductase and it formally occurs by the addition to Pheide a of two atoms of oxygen and four atoms of hydrogen.

The structures of the final products of chlorophyll breakdown, NCC, from different plant species suggest that two further common reactions take place, namely hydroxylation and non-enzymic tautomerization of the fluorescent to the non-fluorescent catabolites.

Breakdown of chlorophyll in higher plants is initiated in the thylakoid membrane and it ends with NCCs located in the central vacuole of senescent mesophyll cells. Hence, catabolism takes place in several subcellular compartments and requires transport processes across membranes. The topographical model is shown in Fig. 3.1. It summarizes the current knowledge of the catabolic pathway of chlorophyll as organized within a degreening mesophyll cell.

The enzymes chlorophyllase and PaO (probably Mg dechelatase) are located in the inner envelope membrane, where as RCC reductase is a soluble protein of the stroma. It has been speculated that a specific transport protein for chlorophyll or plastoglobules plays a role in shuttling pigment molecules between thylakoids and the chloroplast envelope (Thomas, 1997). The factor "X" appears to act upstream of chlorophyllase and is newly synthesized as chlorophyll breakdown is induced. The nature of "X" is unknown. Recently, a water – soluble apoprotein of chlorophyll has been cloned and characterized from cauliflower. The purified protein was able to remove chlorophyll from pigment protein complexes (Satoh *et al.,* 1998). The protein is located in plastids and is a candidate of a putative shuttle for chlorophyll from thylakoids to the site of chlorophyllase.

The final catabolites of chlorophyll breakdown are localized in the vacuole of senescent cells (Hinder *et al.,* 1996). A specific carrier of the tonoplast is responsible for transport across the tonoplast have been identified as members of the family of ATP binding cassette (ABC) transporters. The entire metabolism of catabolites including modifications and conjugations takes place at the level of FCCs prior to transport into vacuoles. The disposal of NCCs in the vacuole is reminiscent of the stepwise detoxification in plant cells of xenobiotics and herbicides. It is reasonable to state that chlorophyll is detoxified rather than metabolized during leaf senescence.

Some questions on different aspects of chlorophyll breakdown remain unanswered. The factor (s) initiating and the mechanisms regulating chlorophyll degradation are elusive. The isolation of the genes encoding the catabolic enzymes and the characterization of their respective promoters will be necessary for addressing these questions.

Fig. 3.2: Topographical model of the pathway of Chl breakdown involving several cellular components. Characterized enzymic reactions and transporters are numbered. Question marks indicate proposed but as yet unknown reactions. An unidentified factor (X) is responsible for the removal of Chl from the thylakoids (after Hortensteiner, 1999). Chl stability in such plants is associated with the persistence of Chl-binding proteins in the thylakoids (Thomas and Hilditch, 1987; Guiamet *et al.,* 1991; Cheung *et al.,* 1993; Bachmann *et al.,* 1994). Apoproteins seem to be protected from proteolysis as long as they are properly complexed with pigments. Since the proteins of pigment complexes in the thylakoids account for over 30% of the total protein of chloroplasts, their recycling contributes substantially to the overall N budget. When *F. pratensis* plants are cultivated under nutritional conditions that promote internal N recycling, the stay-green line pays a significant penalty compared with the wild type in terms of rates of growth and development, largely because thylakoid protein N is immobile in senescent leaves. It would be expected therefore, that stay-green genotypes might be less competitive under natural conditions of N limitation than are wild types with a more efficient internal N economy. It is no wonder then that stay-greens are found above all in cultivated plant species, particularly in legumes such as soybean, French bean, and pea (Thomas and Smart, 1993).

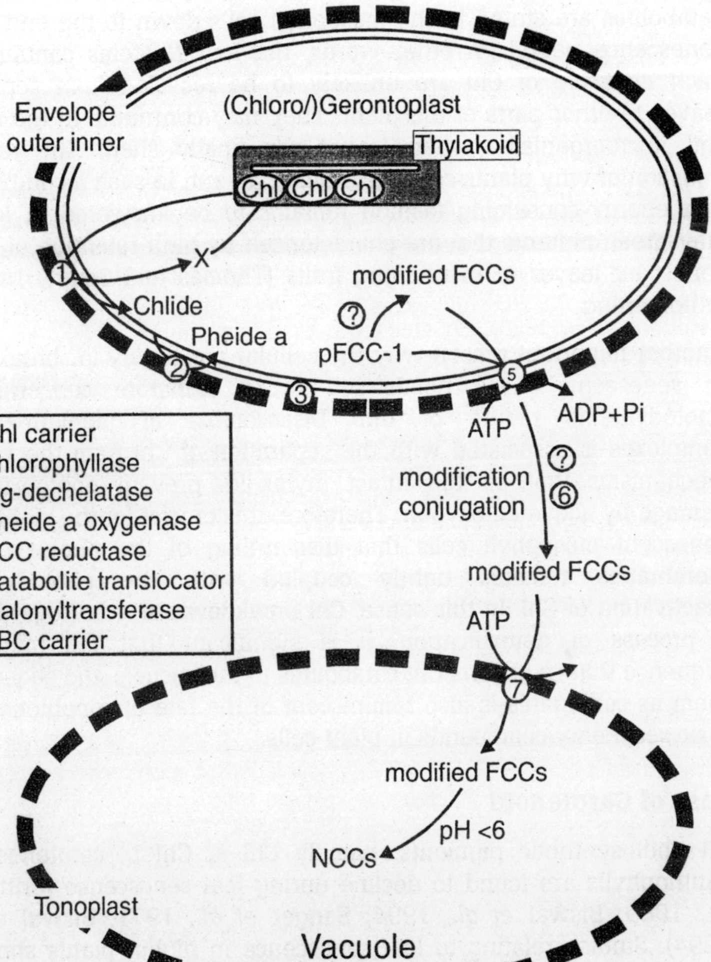
Envelope
outer inner
(Chloro/)Gerontoplast
Thylakoid
Chl Chl Chl
X
modified FCCs
Chlide
Pheide a
pFCC-1
ADP+Pi
ATP
modification
conjugation
modified FCCs
ATP
modified FCCs
pH <6
NCCs
Tonoplast
Vacuole
X Chl carrier
1 Chlorophyllase
2 Mg-dechelatase
3 Pheide a oxygenase
4 RCC reductase
5 Catabolite translocator
6 Malonyltransferase
7 ABC carrier

Significance of Chl Breakdown

If it is justifiable to extrapolate from barley, rape, and a few other species to higher plants in general, then it can be stated that Chl is not catabolized beyond the level of linear tetrapyrroles and that these catabolites are stored in the mesophyll cells down to the end of the senescence period. In other words, the four N atoms contained in each molecule of Chl are unlikely to be recycled from senescent leaves to other parts of the plant. They may contribute to the diet of soil microorganisms when leaves are finally shed. Ant yet, the question of why plants catabolize Chl at all and in such a complicated and energy-consuming fashion remains to be answered. A look at stay-green mutants that are characterized by high retention of Chl in senescent leaves or in ripening fruits (Thomas and Smart, 1993) is enlightening.

Another important reason why the cellular machinery for breakdown in senescent leaves needs to be so elaborate concerns the photodynamic nature of Chl. Disassembly of pigment-protein complexes is associated with the separation of Chl from the various mechanisms that, in the intact thylakoid, prevent photooxidative damage by activated oxygen. Therefore, it is crucial for the viability of senescent mesophyll cells that dismantling of the photosynthetic membranes remains tightly coupled with the photodynamic inactivation of Chl. In this sense, Chl breakdown may be regarded as a process of detoxification. It is significant that the metabolic sequence that transports Chl catabolites to the vacuole and sequesters them as conjugates is also reminiscent of the fate of xenobiotics and toxic secondary compounds in plant cells.

Loss of Carotenoid

All photosynthetic pigments, namely Chl a, Chl b, carotenes and xanthophylls are found to decline during leaf senescence [Britton *et al.,* 1989; Biswal *et al.,* 1994; Sanger *et al.,* 1971; Biswal *et al.,* 1994). Studies relating to leaf senescence in higher plants shows a greater loss of Chls than carotenoids, which results in yellow colouration of senescing leaves. Relative stability of carotenoids is demonstrated during leaf senescence in gymnosperms (Toyama, 1965; Matile, 1994) and angiospermic plants (Biswal *et al.,* 1983, Biswal *et al.,* 1976, Joshi *et al.,* 1991).

Chloroplasts produced during leaf senescence (known as gerontoplasts) are characteristically different from chloroplasts of green mature leaves, the transformation is expected to bring changes in the composition and distribution of carotenoids.

Parallel to a quantitative loss of carotenoids, some of the remaining carotenoids in gerontoplasts turn out to be qualitatively different. Leaf senescence causes formation of xanthophylls acyl esters [Cardini, 1983; Tevini *et al.,* 1985; Britton *et al.,* 1989; Czezuga, 1986; Lichtenthaler, 1969a; Lichtenthaler, 1969b. In Barley a significant increase in the esters with a concomitant decrease in the concentration of free xanthophylls has clearly been demonstrated during barley leaf senescence (Britton *et al.,* 1989). In addition to the formation of esters, there is also a significant change in the composition of xanthophylls. Among the xanthophylls examined so far, lutein, neoxanthin, violaxanthin and anther axanthin decline during senescence [Britton *et al.,* 1989; Biswal *et al.,* 1994; Sanger *et al.,* 1971; Whitifield, 1974). Lutein has been shown to remain remarkably stable compared to other xanthophylls in most of these studies. In fact, a relative increase in the level of lutein is observed during senescence in the leaves of *Festuca pratensis* (Biswal *et al.,* 1994). Senescence-induced qualitative alteration of xanthophylls may also occur, leading to the appearance of other carotenoid species. The possibility of senescence-induced formation of B-carotene epoxide has been discussed by Britton and Young (1989). The epoxide is known to be a product of photo-oxidative events associated with oxidation of β-carotene (Young *et al.,* 1989). Although its presence could not be traced in dark-induced senescence of detached barley leaves (Britton and Young, 1989). Biswal et al. (1994) have demonstrated its appearance during an advanced stage of senescence in the leaves of *Festuca pratensis.*

Thylakoid breakdown is accompanied by the formation of plastogloubuli which increase in number and size with the progress of which are found to contain thylakoid breakdown products including carotenoid esters and free carotenoids such as β-carotene and xanthophylls during senescence (Biswal *et al.,* 1988). Tevini and Steinmuller (1985) have examined the formation of plastoglobuli in relation to deposition of thylakoid breakdown products, including free carotenoids and their esters, in them in detail during the course of senescence of beech leaves in autumn. In non senescing leaves, plastoglobuli are demonstrated to contain small amounts of carotenoids but the pigment exhibits an increase in its content during senescence. With the progress of senescence, there is an increasing tendency towards the appearance of carotenoid esters which constitute the major component of total carotenoids when the leaves turn completely yellow. A similar kind of deposition of caroptenoid esters and free carotenoids during leaf senescence of spinach was earlier demonstrated by Licthenthaler (1969a, b). During advanced stages of senescence, there is a significant increase in the amount of free fatty acids released from triacyl glycerols by the action of hydrolyzing lipases (Biswal *et al.,* 1994). It is possible that these fatty acids form esters with free carotenoids released as the consequence of thylakoid breakdown and which are located in plastoglobuli. Although the precise physiological significance of these carotenoids located in the plastoglobuli is not known, the propostition is that after a significant loss of Chl, the transformed gerontoplasts with more plastoglobuli than thylakoids take over the functions of chloroplasts although in a different way (Schmidt, 1988).

Schmidt (1988) has proposed that light absorbed by carotenoids in plastoglobuli in the absence of Chls may provide the heat required for effective transpiration by leaves and for other cellular metabolic processes particularly in autumnal senescing leaves in temperate climatic conditions. The other role of carotenoids in gerontoplasts is possibly to provide stability to LHCP, which has been shown to remain quite stable despite loss of significant amounts of total Chl. The stability of LHCP may be correlated with the continuation of thylakoid photochemistry even under senescing conditions (Schmidt, 1988). It therefore appears that in addition to the production of heat, carotenoids play a crucial role in residual photosynthesis in gerontoplasts to generate ATP molecules for ordered degradation of cellular structure during leaf senescence (Biswal *et al.,* 1994)

Mechanism of carotenoid degradation

Definite mechanism of carotenoid degradation is not known. Although the enzymes of carotenoid biosynthetic pathways are well characterized, and their genes are even cloned (Hundle *et al.,* 1991; Hundle *et al.,* 1991).

Senescence-induced formation of free radicals and loss of carotenoids

An integrated scheme of free radical medicated degradation of carotenoids and formation of the esters.

Senescence-induced loss in thylakoid photochemistry (Biswal *et al.*, 1994) and a decline in Chl *a* fluorescence (Biswal *et al.*, 1979, Joshi *et al.*, 1993) may lead to the overloading of the quanta on the pigments of chloroplasts. Because of the poor scavenging potential of the organelle owing to loss in the activity of scavenging enzymes during this process (Biswal *et al.*, 1994; Joshi *et al.*, 1993), there is a high probability of production of an enhanced level of toxic oxygen radicals including singlet oxygen ($^1O_2$). Free radicals such as superoxide (O_2) may be formed even at an initial phase of leaf senescence when there is still a significant level of electron transport in thylakoids without any measurable activity of RuDP carboxylase (Biswal *et al.*, 1994) Since CO_2 is the ultimate sink for electrons generated in light reactions in chloroplasts, a loss in calvin cycle efficiency may favour O_2 receiving the electrons leading to formation of $O_2^{\cdot-}$, which may subsequently be oxidized to $^1O_2$ by strong oxidants accumulated at the donor side of PS II due to senescence-induced loss in O_2 evolution ((Biswal *et al.*, 1990). The $^1O_2$ can also be produced by triplet Chl (^{3}Chl*) which is known to be formed because of alteration in thylakoid structure and release of free Chl during senescence. These free Chls and their breakdown products are prone to triplet formation (Asada *et al.*, 1987). Both ^{3}Chl* and $^1O_2$ can directly oxidize carotenoids (Takahama, 1983) or it can indirectly contribute to their degradation. The oxygen radicals may result in breakdown of thylakoid lipids liberating free fatty acids (Biswal *et al.*, 1994) which subsequently from esters with xanthophylls. In addition to the formation of esters, free fatty acids upon release may also act as natural detergents and disassemble the components including carotenoids from pigment-protein complexes of thylakoids. The free carotenoids may subsequently become prone to enzymatic and /or non enzymatic degradation.

Enzymatic degradation

Senescence is an integral part of plant development. The following experimental findings support enzyme participation for its degradation

Dark-induced senescence of attached and detached leaves results in loss of both Chl and carotenoids (Biswal *et al.,* 1976; Joshi *et al.,* 1991. Biswal *et al.,* 1983; Mishra *et al.,* 1981). In continuous darkness the participation of free radicals in destroying carotenoid molecules is minimized and its degradation may, therefore, be attributed to the involvement of enzymes. This is further supported by the findings of Tetley and Thimann (1974), who have demonstrated degradation of carotenoids in etiolated seedlings grown in the dark even with the same rate of degradation of Chl in green leaves. Since phytohormones are known to act at the level of enzymes, further evidence supporting the involvement of enzymes in the breakdown of carotenoids comes from their results on kinetin –induced retardation of the pigment loss. A similar retarding effect of the hormone on carotenoid loss has been reported by Biswal and Mohanty during senescence of barley leaves (1976).

The data on carotenoid metabolism during leaf senescence of non-yellowing mutants of *Festuca pratensis* also favour enzymatic degradation of the pigment. Thomas and Matile (1988) have demonstrated almost full retention of the photosynthetic pigments in the mutant despite complete loss of the scavenging system and protective mechanism against photo-oxidative destruction, which would mean involvement of a process other than photo-oxidation for degradation of the pigment. Since the mutation of a *unclear gene* in *Festuca* results in high retention of carotenoids, the control of carotenoid degradation by genes through enzyme system (s) is suggested.

The proposition that carotenoids degrade independent of action of oxygen free radicals in supported by the observations that light, which is known to generate these radicals, rather retards the pigment loss during leaf senescence of higher plants (Biswal et al, 1976; Joshi *et al.,* 1991) and ferns (Whitified, 1974).

1. While searching for the location and nature of enzymes (Friend, 1960) and Friend and Acton (1966) have done extensive experimentation with chloroplasts of sugar beet leaves. In their experiments using water soluble carotenoid glycoside crocin as a model carotenoid, they have demonstrated degradation of crocin by isolated chloroplasts. While examining the response of different extracts of chloroplasts to various additives including inhibitors, metal salts and metal chelating agents on crocin

oxidation, they suggested the involvement of a haematin type of copper protein enzyme that could possible be responsible for pigment catabolism. The possibility of a peroxidase-catalyzing carotenoid degradation has been suggested by Matile and Martinoia (1982). They have found peroxidase activity in the thylakoid preparation which could efficiently bleach lutein, a major xanthophylls in higher plants. These in vitro experiments with thylakoid preparation indicate two things: that enzymes participate in carotenoid breakdown: and that these enzymes are located in chloroplasts.

2. However, in spite of these findings the exact nature of the enzyme (s) largely remains unclear. Review of the literature indicates that is most likely that both enzymatic and photoxidative events are operating during senescence, each contributing to the total loss of carotenoids. It is also possible that both the photo-oxidative process and the enzymatic participation are the two components of a single mechanism operating for degradation of the pigment.

Carotenoid degradation: a coordinated event

Chl, carotenoids and leaf proteins exhibit simultaneous loss during induction of senescence (Biswal *et al.,* 1976; Joshi *et al.,* 1991; Biswal., 1983). Although the loss of proteins and Chl is relatively rapid compared to the loss of carotenoids the general kinetic patterns of loss in all these cases remain almost the same. The pattern of retardation of loss of both these pigments by different senescence modulators including phytohormones also remains more or less the same (Biswal *et al.,* 1976; Joshi *et al.,* 1991; Mishra *et al.,* 1981). Retention of Chl in non-yellowing mutants of *Festuca pratensis* is also accompanied by a similar retention of carotenoids. These findings may suggest a coordinated breakdown of photosynthetic pigments during senescence. (Gut *et al.,* 1987)

Since carotenoids are part of different Chl-protein complexes, the stability of these complexes may contribute to relative stability of the pigments. An increase and decrease in Chl a/b suggests the relative stability of RC II complexes and LHCP respectively. Since chl b is associated only with the light-harvesting complex of green plants and Chl a is associated with this complex in addition to its binding with reaction centres of photosystems. The value of the Ch a/b ratio is taken as an index of relative stability of LHCP and reaction centres of photosystems, the relative stability of LHCP and reaction centres of photosystems. The relative stability of Chl b during senescence, and a decline in Chl a/b ratio as observed by various workers in different plant systems during dark-induced or natural senescence may therefore suggest stability of LHCP, and this could be linked to high retention of lutein during senescence (Britton *et al.*, 1989; Biswal *et al.*, 1994; Sanger *et al.*, 1971). Similarly, the relative stability of the reaction centre complex of photosystems may be correlated with retention of β-carotene (Biswal *et al.*, 1994; Sanger *et al.*, 1971; Gut *et al.*, 1987), Lutein and β-carotene are known to be predominantly located in LHCP and reaction centre complex respectively. These carotenoids may remain stable as long as these complexes remain intact during senescence. Violaxanthin is known to be the most unstable carotenoid species during leaf senescence (Sanger *et al.*, 1971). This xanthophylls is believed to be distributed throughout the thylakoid surface and therefore becomes susceptible to senescence-induced degradation by any non-specific changes associated with the photosynthetic organelle.

Whatever the location of carotenoids, since these pigments exist in the form of complexes with proteins, Chl and presumably in association with specific lipids, breakdown of any individual component may lead to the collapse of the complex. Disassembly of the complex is in fact the prerequisite for enzymatic degradation of individual components including carotenoids. However, what is the component that degrades first during senescence, is still to be answered.

Possible operation of the xanthophylls cycle during senescence

When green plants receive more light energy than is necessary for photosynthesis, the excess energy results in the damage of photosynthetic apparatus, a process known as photoinhibition. Although the process is very complex and is still under active investigation, the target site of photoinhibition is known to be PS II of thylakoids (Demming-Adams, 1992). Photoinhibition is controlled by many factors and the plants have developed their own strategy to counter the inhibiting action of excess light. One of the protective mechanisms against photoinhibition is the operation of the xanthophylls cycle, which has been widely examined in various plant systems in different environmental conditions (Demming-Adams, 1992; Demming-Adams, 1995; Adams, W. W. III, 1994). The cycle involves conversion of xanthophylls, namely violaxanthin, to zeaxanthin and vice versa, depending on the amount of light quanta the plants receive during irradiation.

In cyanobacteria, although the viloaxanthin / antheraxanthin / zeaxanthin cycle does not operate, the stress causes the formation of zeaxanthin (Biswal *et al.*, 1994) and the xanthophylls is known to be formed through the β-carotene / β-cryptoxanthin / zeaxanthin route (Demming-Adams, 1990).

Since some of the basic photoinhibitory features and other stress responses such as loss in PS II photoelectron transport, changes in fluorescence properties of thylakoids, formation of free radicals, thylakoids, formation of free radicals, and thylakoid membrane lipid peroxidation are also demonstrated to be associated with senescing leaves, the following question arises: is the xanthophylls cycle operating during senescence too? Experiments conducted so far in different plant systems do not suggest operation of the cycle during the process. An apparent decrease in neoxanthin in leaves of *Meadow fescue* (Biswal and Young, 1989) and also a significant decrease in neoxanthin, violaxanthin and lutein in barley leaves (Britton *et al.*, 1989) do not result in the formation of any significant increase of zeaxanthin during senescence. Therefore, the loss of violaxanthin and other xanthophylls during senescence may not necessarily be indicative of operation of the xanthophylls cycle. Secondly, there is no evidence of any increase in the pool size of xanthophylls cycle components during senescence in these systems. During cotyledonary senescence of clusterbean, there is no significant enhancement in the level of zeaxanthin. However, high retention of β-carotene has been shown to stabilize D1 protein of thylakoids during senescence. It appears that the xanthophylls cycle is inoperative during leaf senescence. A substantial decrease in the level of violaxanthin, neoxanthin, antheraxanthin and other xanthophylls induced by senescence has a correlation with the formation of their acyl esters (Britton *et al.*, 1989) instead of their routes through the xanthophylls cycle. In fact, operation of the xanthophylls cycle may not be a general response of high light and other stress conditions. Recently Senger et al. (1993) have shown operation of the β-carotene-lutein-loroxanthin pathway rather than the operation of the xanthophylls cycle in the green alga *Scenedesmus* under different light conditions.

In this connection, the results of relative stability of lutein compared to other carotenoids as noted by many authors during senescence remain to be explained.

It is not very difficult to explain the reason for, and logic of, the incorporation of the xanthophylls cycle during leaf senescence. In spite of many similarities, senescence (a physiological process) is different from photoinhibition (basically a stress response). The formation of zeaxanthin needs activation of de-epoxidase enzyme in a highly acidic environment of thylakoid lumen and this condition is provided by high light irradiation. Senescence-induced drastic decline in electron transport efficiency of thylakoids (Biswal *et al.,* 1988) however, may not result in the formation of the level of proton gradient that could enhance lumen acidity for activation of the enzyme. This may be the major factor for absence of the cycle during senescence. Absence of operation of the cycle during senescence, however, has an important physiological significance. Since during leaf senescence most of the Chls are lost and residual light-absorbing pigments have to absorb light to maintain a basal level of photosynthesis provide energy for breakdown of cellular components, formation of zeaxanthin may dissipate these limited light quanta thus further lowering the capacity of senescing leaves for photosynthesis. Absence of any significant amount of zeaxanthin during senescence, therefore, does not permit loss of this energy.

LOSS OF LIPIDS AND MEMBRANES

The outstanding change, which occurs in the cells of senescing plant, is the deterioration of membranes. Changes in cell's membrane composition and membrane permeability occur during senescence. Several studies have been conducted on changes in the ultra structure of leaves and cotyledons during senescence. In photosynthetic tissue, the chloroplast thylakoids are typically the first membrane to show signs of deterioration. As deterioration of thylakoid continues, the ER, mitochondria, and tonoplast show evidence of deterioration. Such changes also occur in non-photosynthetic tissues like cotyledons, roots, xylem, and phloem during senescence. In photosynthetic tissue, lipids associated with the chloroplast are the first lipids to decline, while the non-chloroplast lipids decline later. In plasma, microsomal, and chloroplast membranes, it has been reported that there are 400% increase in the sterol: phospholipid ratio as senescence progresses.

Changes in fatty acid content have also been reported in many tissues undergoing senescence. Degradation of lipid and the subsequent action of lipoxygenase on fatty acids play an important role in destruction of the various membranes of the cell. The activity of superoxide dismutase decreases during senescence. Leshem (1984) proposed a model for the degradation of membrane based upon acyl hydrolase and lipoxygenase activities.

Many studies reveal that membranes in mature tissue are in liquid crystalline state, while in senescent tissue of the membranes, or portion thereof, are in crystalline gel phase at physiological temperature. Therefore, during senescence the membrane lipid progressively becomes crystalline. Senescence process increases the permeability of the membrane and so contributes to loss of compartmentation. It is very likely that the increase in permeability is at least partial due to the occurrence of the crystalline gel phase in membrane lipids. It also helps in translocation of materials from cotyledons to other tissue.

X-ray diffraction studies show structural changes of chloroplast and microsomal membranes during leaf senescence. (McKersie and Thompson, 1978; Chia, 1981). As the tissue senesces, the membrane lipids progressively become crystalline. The structural change seems to be associated with a 4-fold increase in the sterol to phospholipid ratio, but it seems to be independent of the change in the ratio of the different fatty acids components of phospholipids. Senescent mesophyll cells usually accumulate lipid granules in chloroplast (Cohen *et al.,* 1979; Wilhelmova, 1981), which, probably derives from thylakoid breakdown. In cytoplasm of senescent mesophyll cells also accumulates lipid grain or oleosomes. (Parker and Murphy, 1981) Both malonyldialdehyde (Chia *et al.,* 1981) and lipid peroxides (Dhindsa *et al.,* 1981) increase in senescence. In attached tobacco leaves, the increase in lipid peroxides is parallel with the decrease of chlorophyll and protein. Other lipid changes associated with senescence is differential rates of degradation of different types of membrane lipids, namely monogalactosyldiglycerides, digalactosyldiglycerides, sulfoquinvosyl, diglycerids, phosphatidyl glycerol, phosphatidyl choline and phosphatidyl inositol. (Tevini and Steinmuller, 1985) The degradation of lipids is attributed to senescence-induced stimulation in the activity of lipases (Woolhouse, 1984). It is important to note that this acid greatly increases during chloroplast biogenesis.

As tissue senescence, the membrane becomes crystalline. For this, it has been suggested (Dhindsa *et al.,* 1981) that free radical induced peroxidative reactions could modify the membrane lipids and triglycerides. Lipid peroxides may decompose to malonyl-dialdehyde, which attacks various cellular components (Wilhelm and Wilhelmova, 1981). Lipid peroxides probably also transform into other lipid derivatives, which cannot be metabolized determinating structural modification in membranes. Dhindsa *et al.,* (1981) attribute the increase of lipid peroxides to a previous increase in the concentration of the radical superoxide (O_2) due to a diminution of the enzymes superoxide dismutase and catalase, which destroy O_2. According to this interpretation, lipid peroxidation is a consequence of the general decrease of proteins (among them superoxide dismutase and catalase), which occurs in senescence. The change in the properties of the membranes during senescence may be important if they facilitate withdraw of the components of the cellular compartment. However, it is difficult to find a role in senescence for the accumulation of lipid granules. Senescence results in degradation of membrane lipids, namely monogalactosyl, diglycerides, galtosyl diglycerides, sulfoquinosyl diglycerides, phosphotidyl glycerol, phosphatidyl choline, phosphatidyl inositol. Here arises a question that is this symptom a primary factor for controlling other posterior symptoms? Related to the decrease of galactolipids, Dalgarn et al. (1979) found a decrease in galactosyl transferase of chloroplast envelopes during senescence of soybean cotyledons, but hormonal studies on these aspects are still not available. During soybean cotyledons the percentage of linolenic acid decrease. One of the most rapid effects of zeatin (2 hr) has been described for its stimulation of the increase of linolenic acid in the leaves of *Coleus blumei* and *Impatines sultani*. Rapid effects of kinetin have also been described for the synthesis of lipids by mesophyll protoplasts (Kull and Ultes, 1980). The possible relation of these last results with senescence is not clear. On the one hand, zeatin was applied at very high concentration (100μg/mL). On the other hand, the results may reveal a general effects of cytokinins are not necessary implicated in senescence. Its importance would depend on the membrane lipids turnover during senescence and in mature leaf. However, the results show the importance of cytokinin action on membrane composition. It is possible that appropriate membrane integrity and composition depend on a continuous supply of cytokinin to the leaf. A diminution in the level of cytokinin at the beginning of senescence may determine membrane changes. The relation of these changes in linolenic or in other fatty acids with the membrane changes in senescence in still unclear.

LOSS OF MACROMOLECULES

Nucleic acids

The characteristic decrease in protein content in senescing leaves is accompanied by a parallel decline in RNA content. This decline in RNA content is associated with a reduced capacity of senescing leaves to incorporate radioactive precursor into nucleic acids. In contrast, leaves treated with growth regulators which retard senescence the capacity to utilize precursor is sustained. This evidence suggests that decline in RNA content during senescence is due to the decomposing capacity for RNA synthesis and it has been suggested that senescence is caused by failure of DNA to provide an effective templates for RNA synthesis.

If senescence is associated with decomposition in DNA template, availability might be expected that there would be a reduction in the synthesis of RNA components in order to detect changes in mRNA. This method for determining relative levels at polysomes to monosomes in ribosomal preparation has been used. Ribosomal preparation has been obtained from senescing leaves which, showed that though the polysomes level declines during the growth of the leaf, the mature leaf retain cytoplasmic polysomes but chloroplast polysomes are absent only in younger leaves. The absence of cytoplasmic polysomes at later stages of senescence may indicate that mRNA is long lived or alternative mRNA synthesis may be sustained even when the leaves are yellowing and senescing. The mRNA synthesized now could be loading for the production of hydrolytic enzymes during senescence although polysomes may still be absent.

The total ribosome content of leaves declines during senescence. In initial stages, the properties of chloroplast ribosome declines, later on free cytoplasmic ribosomes are depleted whereas the membrane bound ribosomes are sustained during senescence there is an increase in the rate of soluble RNA to rRNA suggesting a greater liability or reduced synthesis of rRNA. However, the evidence supporting this concept is extremely conflicting.

In any event, whether causative or not, there are marked changes in the nucleic acids of senescing leaves (Goldney and Van Steveninck, 1972). A general decline in total RNA was first noted by Smillie and Krotokov (1961) in pea leaves, in which both RNA and DNA decrease steadily from very early stages. Similar decreases, specifically during senescence, were reported by Bottger and Wollgiehn (1970) in tobacco leaves, and later, using varied plant species, by Osborne et al. (1962) Shaw et al. (1965), Fletcher et al. (1965), and Baddley et al. (1969). The decrease in DNA is typically smaller than that in RNA. Wollgiehn found with tobacco leaves that after 3 days the RNA had decreased by 16 per cent and DNA by 3 per cent. Osborne (1962) with *Xanthium* leaves found that after 4 days the RNA had decrease by 55 per cent and the DNA by 39 per cent. While total RNA decreases, the synthesis of some RNA continues. It is found that incorporation of adenine or orotic acid into the RNA of chloroplast, mitochondria, ribosomes, and cytoplasm is still continuing, though slowly, even when the leaves are fully yellow, Baddley et al. (1969) noted that cytoplasmic r-RNA continues to be synthesized for a time in senescing wheat leaves. Synthesis and breakdown of protein are similarly continuing together, as noted above. The loss of RNA can be due to decrease in RNA polymerase (Schneider, 1980) or due to increase in activity of RNAase.

Loss of RNA can be due to decreases in the rates of synthesis or to increases in the activity of RNAse. The latter increases can be large, but our knowledge of them is imperfect. In senescing corn leaves the RNAse level became high enough that it was considered a very rich source of the enzyme (Kisaki, Hirabayashi and Yano, 1973). In oat and barley leaves, however, the total RNAse rises at first and then falls again to its initial level on about the 4th day (Inada, 1967). However, the situation is made more difficult by the presence of more than one RNAse in most leaves. For instance, only one of the four nucleases in oat seedlings was found to increase during senescence (Carlson, 1973). Similarly, senescing barley leaves showed no marked change in total RNA in 4 days, but when the chromatin was isolated from these leaves both the DNAase and RNAase associated with it had increased some 10 fold (Inada, 1967; Incoll and Neales, 1970). The decrease in RNA is of course accompanied by disappearance of ribosomes, which indeed seems to start as soon as growth stops (Baddeley and Simon, 1969). The decrease in DNA is typically smaller than that in RNA. The reason for this is that DNA has to control various biochemical processes until at the end of plant dies off.

Proteins/Amino acid

The protein molecules in the plants are in a constant state of turnover and at steady state condition when protein content remains constant, the rate of synthesis is equal to the rate of degradation. In leaves of annual plants, the protein content does not remain constant for a long period. Instead the leaf protein content reaches high at the time, when the leaf becomes fully expanded and thereafter it begins to decline. The initial decline in leaf protein occurs before there are any visible indications of chlorophyll depletion or other outward signs of the onset of senescence. The decline in protein content and onset of senescence are accelerated by leaf attachment. It was observed that the capacity to incorporate exogenously supplied labeled amino acids decline as detached leaves are immersed in water. However, when these leaves are treated with growth regulators (which retard senescence) the ability to incorporate amino acids into protein a retained. This led scientists to conclude that senescence caused by a failure or a declining capacity is protein synthesis.

During senescence, amino acids are accumulated in detached leaves and with time there is an increase utilization of these amino acids as respiratory substrates, while the carbon skeletons are used in respiratory metabolism. The nitrogen components appear to be accumulated in the amino residues of glutamine and aspartate. In early senescence, glutamine formation predominates while in the later stages of senescence aspartate biosynthesis predominates.

Kao (1981) observed a striking increase in proline content in the senescing leaves of rice. He noticed that, first there is a slight but significant increase in the proline content during first day of dark incubation, but subsequently proline accumulation becomes relatively rapid and after 4 days of excision, 30 to 50 folds increase the proline level in the leaves. This proline accumulation during senescence was observed in many other plants besides rice leaves.

Proteins breakdown during leaf senescence has been realized from some of the earliest studies. Yemm Laloue, (1977), Lamarche (1969) made comparable observations on detached barley leaves and noted that the respiratory quotient fell after the first day to around 0.75, which suggested that the amino acids liberated by proteolysis which is the main respiratory substrate. Kennedy and Laetsch (1973) also showed extensive protein breakdown. Since then virtually all workers, using widely different plant species and different experimental conditions, and noted protein hydrolysis (Wood and Cruikshank (1944), Osborne (1960,64), Wollgiehn (1961), Eklund and Moore (1969), Fletcher (1965), Shibaoka and Thimann (1973), Parisco and O'Leary (1972), Martin and Thimann (1974), and Tetley and Thimann (1962). Wittenbach (1982) working on isolated wheat protoplast found that there is loss of protein content. The amino acids are produced due to the degradation of proteins, which migrates to young or to reproductive structures. Decline in protein in the chloroplast involves a decrease in the major chloroplast enzyme Rubisco (Callow, 1974; Kannangara and Woolhouse, 1967; Kawashimai *et al.,* 1967) along with other enzymes known to be synthesize at least in part on chloroplast ribosomes (Batt and Woolhouse, 1975; Spencer and Titus, 1972). Wittenbach et al. (1982) working on isolated chloroplast of wheat found that there is decrease in Rubisco. A typical symptom of senescence of attached leaves in natural condition is the decrease of free amino acid in leaves (Story and Beevers, 1978; Kao, 1980; Kang and Titus, 1980) The decrease is usually interpreted as due to the rapid transport of the products of proteolysis from the senescent leaves to young or to reproductive structures. In contrast, in detached leaves free amino acids accumulate during senescence (Kao, 1980; Mishra, 1965), which is interpreted as due to the absence of sink structures for the amino acids in this system. In attached leaves of peas (Storey and Beevers, 1978) the decrease of free amino acids levels is parallel with chlorophyll decrease and slightly precedes protein decrease.

Protein degradation determines, or at least is parallel, with a diminution of the photosynthetic capacity of the leaf and with ability of the leaf to export carbohydrates. Carbohydrates are also required as energy source and as carbon precursors for biosynthesis in reproductive structures. One can think that an adjustment of the nitrogen and carbohydrate demands of developing reproductive structures may control leaf senescence in monocarpic plants, but the problem may be difficult for an experimental examination. In fact, frequently reproductive structures can carry our considerable photosynthesis. Bazzaz et al. (1979) estimated that in a number of plants the probable contribution of photosynthesis of reproductive structures to their total carbon demands are in the range of 50%.

Ofcourse, this contribution varies along the reproductive structure and with the senescence of the photosynthetic apparatus. At the beginning of senescence in *Lolium temulentum* leaves, Lloyd (1980) found a rapid loss of insoluble leaf material (probably including polysaccharide previously marked with [^{14}C]-sucrose or $^{14}CO_2$. As soon as 6 hr after the induction of senescence, an increase of proteolysis is detected in oat leaves (Shibaok and Thimann, 1970) and polysaccharide breakdown soon follows. While studying grain filling in different oat varieties, Welch et al. (1979) found that the relative low non-structural carbohydrate content of the high grain protein varieties is apparent only at the end of the period of grain filling. However, grain protein content differences among varieties are apparent during all the period of grain filling.

The high protein varieties have more efficient mechanisms for the mobilization of nitrogen and its translocation to the grains. They attribute the low carbohydrate contents of the high grain protein varieties to an increased level of respiration associated with the greater mobilization and translocation of nitrogen. However, a low photosynthesis cannot be excluded in the protein rich varieties at the end of the period of grain filling. A typical symptom of senescence of attached leaves in natural condition is the decrease of free amino acids in leaves (Story and Beevers, 1978; Kao, 1980b; Kang and Titus, 1980).

It is not clear if this is a general phenomenon, but it's knowledge would be of interest in relation with the mechanism of senescence. In contrast with the senescence in natural conditions, if senescence is induced by exposure of attached leaves to dark, free amino acids increase at the beginning of senescence (Wittenbach, 1978; Lloyd, 1980) in parallel with a decrease of soluble protein. The relation of this dark induced senescence with the nature is uncertain. But it must be noted that usually senescence of detached leaves is carried out in the dark, which means that the increase of free amino acids in the senescence of detached leaves may not be due to the absence of sink structures. One possibility is that in dark induced senescence in attached leaves, there is little sink activity of the young structures. Indeed, Mor and Halevy (1980a) found that sink activity of young shoot is promoted by light. However, this light promoted activity of young structures seems to be mediated through phytochrome (Mor and Halevy, 1980b) but in *Lolium* (Lloyd, 1980) senescence is also induced by CO_2 starvation at light. This senescence in the absence of

CO_2, although not as quick as the dark induced senescence, is also accompanied by an increase of free amino acids and a decrease of protein in the leaf. Today, the reason of the decrease of amino acids in attached leaves and its increase in detached leaves is not clear. The hypothesis of the presence of sinks in assays with attached leaves is, at least, too simple. The possibility of an effect of the light favoring an export of amino acids through a direct effect in the leaves cannot be excluded. Years ago, Chibnall (1939) described increased in free amino acids in attached leaves in night and fall again in the day.

SENESCENCE INDUCED ULTRASTRUCTURAL CHANGES

Senescence-induced ultrastructural modifications (including changes in the molecular composition of thylakoids) have been probed by electron microscopy, X-ray diffraction, immunotechniques, absorption and fluorescence spectroscopy, and phase properties of membranes. Electron microscopy has been used extensively by many authors to examine the changes in size, shape and ultrastructural organization of chloroplasts during senescence (Butler and Simon, 1971; Thomas, 1977; Naito *et al.,* 1981; Wittenbach *et al.,* 1982; Biswal *et al.,* 1983a). Chloroplasts of completely yellow leaves appear oval and exhibit minimal lengths and widths indicating senescence-induced shrinkage of the organelle (Naito *et al.,* 1981). During senescence, the organelle shows disorganization at the level of both the envelope and lamellar structures. Lamellar breakdown, as revealed by electron microscopy, exhibits unstacking of grana thylakoids (Mlodzianowski and Mlodzianowski, 1973). This is followed by the formation of loose, elongated lamellae scattered in the organelle. In the next phase, these lamellae undergo further degradation with the concomitant appearance of plastoglobuli (Hudak, 1981; Biswal *et al.,* 1983a). The size of the globuli increases as senescence advances (Kulandaivelu and Senger, 1976b). Although these ultrastructural modifications have been qualitatively studied, attempts have been made by Naito et al. (1981) to quantify some of the structural parameters. A decrease ca be observed in the number of grana per chloroplast section and in granum length with an increase in the number, length, and width of plastoglobuli during senescence. They have also noted senescence-induced qualitative changes such as the shrinkage of chloroplasts, disappearance of intergranal lamellae, and random orientation of grana followed by their disappearance. Chloroplast degeneration at the final stage of senescence in marked by the disruption of the chloroplast envelope, which leads to rapid lysis of the rest of the chloroplast components (Woolhouse, 1984a).

The sequence of ultrastructural changes during chloroplast senescence varies depending on the environmental conditions and plant species. For example, the pattern of ultrastructural changes of chloroplasts during senescence of naturally occurring attached primary leaves of wheat is qualitatively different from the sequence of changes observed in the chloroplasts on detachment of leaves (Hurkman, 1979). The initial event in the degradation of chloroplasts of attached leaves undergoing natural senescence is the appearance of plastoglobuli and the simultaneous disorganization of the thylakoid system. This is followed by the formation of vesicles and an increase in the number and size of plastoglobuli. In the final stage, the chloroplast envelope ruptures and the plastid contents, namely plastoglobuli and the remains of the thylakoids, fall into the cell lumen. On the other hand, plastid degradation during senescence of wheat leaves detached from the parent plant begins with the rupture of the envelope followed by the collapse of the thylakoid system. In this case, swelling of the grana and intergranal lamellae is not significant and lamellar system breaks down without the formation of vesicles and numerous plastoglobuli. Similar, mustard cotyledons undergoing senescence in different environmental conditions show qualitatively different ultrastructural modifications. Senescent chloroplasts of cotyledons of naturally growing mustard plants exposed to daily light-dark cycles show the formation of giant grana structures (Hudak, 1981) that are not observed in cotyledon chloroplasts during senescence induced by continuous dark treatment (Biswal *et al.,* 1983a). The ultrastructural changes of mustard cotyledons during dark-induced senescence, and the effect of light through phytochromes in retarding these changes can be observed.

The pattern of ultrastructural changes of chloroplasts differs with plant species, even though the experimental conditions remain same. For example, rupture of the plastid envelope is followed by the degradation of the thylakoid system during senescence of wheat leaves (Hurkman, 1979), however in the chicory leaf, the thylakoid system degrades first and is followed by the rupture of the envelope, although in both cases, detached leaves were used for the experiments (Mlodzianowski and Mlodzianowski, 1973).

Ultrastructural degradation consists of three major events: thylakoid breakdown, formation of plastoglobuli, and rupture of the envelope.

The precise mechanism of senescence-induced thylakoid breakdown is not known. Unstacking of grana, the first step of lamellar disorganization, could be due to the loss of chlorophyll b and LHCP (Biswal *et al.,* 1983a). Chlorophyll *b* and LHCP are known to be responsible for grana stacking and both of them are reported to degrade during induced senescence (Lichtenthaler *et al.,* 1981; Biswal *et al.,* 1983a). It is also possible that a senescence-induced loss of ions could result in destacking (Raval *et al.,* 1984). Lamellar degradation the next event of thylakoid break down, could be carried out by the specific enzyme systems. The enzymes for the breakdown of thylakoids are reported to be associated with the organelle itself (Dalling *et al.,* 1983; Thomas, 1984; Wettern and Galling, 1985; Martin *et al.,* 1986).

The breakdown of thylakoid membranes and the concomitant formation of plastoglobuli have led to the proposal that these globuli are composed of lipids resulting from senescence-induced lamellar breakdown. This is further confirmed by the fact that as senescence advances, the increased thylakoid breakdown is accompanied by a corresponding increase in the number of size of the plastoglobuli (Biswal *et al.,* 1983a). The structural closeness of thylakoids and globuli is distinctly shown in senescent chloroplasts of *Phaseolus* leaves (Barton, 1966) in which characteristic swelling of each thylakoid terminus is seen adjacent to dense globuli.

Cytokinin treatment prevents an increase in the number of plastoglobuli and simultaneously prevents thylakoid disintegration (Naito *et al.,* 1981). This also suggests a correlation between the level of plastoglobuli and the structural status of thylakoid membranes. On the other hand, some reports suggest that plastoglobuli may only in part be the product of lamellar breakdown and that the rest of the rest of their structure may represent a nonspecific storage of insoluble lipids (Bailey and Whyborn, 1963; Lewington *et al.,* 1967). This view is supported by the observation that the globuli from mature chloroplasts of *Vicia faba* and spinach contain a mixture of lipids and other substances that are not specific to the lamellar membrane system.

The metabolism of plastoglobuli with reference to thylakoid breakdown during chloroplast senescence was recently investigated by Tevini and Steinmuller (1985). They have analyzed the dynamic changes in the levels of galactolipids, carotenoids, plastoquinones, and other lipids simultaneously in the thylakoids and plastoglobuli and have proposed a correlation between these two structural units. The fatty acids, hydrolysis products of thylakoid lipids, are suggested to form esters of carotenoids in plastoglobuli. In addition to carotenoids and fatty acids, prenylquinones, components of thylakoid, also deposit in plastoglobuli. Glycolipids, the major constituents of thylakoid lipids, do not deposit in plastoglobuli, and therefore the degradation of these lipids, as that of proteins, is insignificant in the metabolism of plastoglobuli. Although these data provide some basic information on the chemical composition of plastoglobuli, the biochemical significance of plastoglobuli during senescence merits further investigation.

The retention of the integrity of the plastid envelope and plasma membrane may suggest the involvement of these membrane system in finally transporting the degradative products to other growing cells in the plants (Woolhouse, 1984a). Rupture of the envelope therefore, represents the last sequence of the structural degeneration of chloroplasts. Two tentative mechanisms can be proposed to explain envelope rupture one of which is mechanical. In this case, the rupture is correlated with the senescence-induced extensive vacuolation of chloroplasts (Hudak, 1981). The origin of vacuoles is attributed to the local lysis of chloroplast stroma. The vacuolation could exert pressure on the chloroplast envelope resulting in its rupture. The second mechanism proposes that the rupture of the envelope may occur by the hydrolytic enzymes released by breakage of tonoplasts. The breakdown of tonoplasts is a crucial phase of senescence because it provides for the direct contact of cellular vacuoles and different cell organelles, including chloroplasts (Peoples *et al.,* 1980). The cytoplasmic vacuoles containing relatively high concentrations of phenols, organic acids, and hydrolytic enzymes could act on the boundary envelope of chloroplasts resulting in its breakdown.

Chapter 4

LEAF SENESCENCE: PHYSIOLOGICAL AND BIOCHEMICAL CHANGES

Senescence in general denotes the onset of a complex series of detrimental changes ultimately leading to the death of a living organism. Leaf senescence is an integral part of the leaf development. It is like other developmental processes- a well-organized and regulated process rather than a passive decay of the leaf. It has been studied and worked upon extensively by a large number of scientists from time to time as these organs form a useful experimental system in which senescence can be studied properly.

Senescence is regarded as the series of event concerned with cellular disassembly in the leaf and the mobilization of materials released during this process. This may also be defined as the resultant of a number of deteriorative processes set in motion according to hereditary instruction, which are aimed at terminating the functional life of an organ or organism under natural condition. Considerable quantities of fatty acids may be liberated during the breakdown of thylakoid lipids, which can be further metabolized in the peroxisomes and finally fuel the tricarboxylic acid cycle in the mitochondria. These suggest that the sequential senescence is the sequential loss of photosynthesis and respiration, a well organized process and not a decay of leaf.

Various biochemical and photochemical changes occur during senescence like photoelectron transport system, oxygen evolving system, photophosphorylation, photosynthetic rate, respiration rate, nutritional status, hormone level, mineral uptake, enzymes, stomatal conductance, seed filling period and yield

Most of the biochemical work deals with detached organs rather than whole plants or attached organs, which may represent a more natural situation. In leaves, and presumably other organs, many components (protein, nucleic acid, chlorophyll, etc.) or process (photosynthesis) may reach a maximum about the time maximum size (growth) is attained; this peak may be followed by an early, slow decline and then a later, rapid decline. The exact significance of the early, slow phase is still uncertain. The early, slow phase could be a metabolic step down following cessation of growth, but there is at present no reason to believe that it is not part of the continuum of degeneration leading to death.

PHOTOELECTRON TRANSPORT SYSTEM

The pigments associated with thylakoid membranes are characteristically assembled into two photosystems, namely photosystem I (PS I) and photosystem II (PS II), that are linked by a series of electron-transport carriers (Cramer *et al.*, 1985). The two photosystems have specific orientations across the thylakoid membrane (Haehnel, 1984). It is proposed that PS I is located in stroma lamellae and is primarily responsible for cyclic electron flow. PS I in grana, on the other hand, is linked with PS II and is largely involved in the reduction of $NADP^+$. Normally, higher wavelength-absorbing pigments are present in PS I that do not provide any significant contribution to the fluorescence emitted by chloroplasts. The reaction center of PS I is known as P700, which is not directly associated with the oxygen-evolving complex. The major electron transport carriers associated with PS I are plastocynin on the oxidizing side, ferredoxin, P430, and NADP reductase on its reducing side. The pigments associated with PS II are the short wavelength form of chlorophyll *a* and other accessory pigments that contribute significantly to the total fluorescence of chloroplasts. PS II is responsible for oxygen evolution, although the oxygen-evolving complex has been identified as a separate structural unit (Raval and Biswal, 1985). P680 is the reaction center of PS II. The major electron-transport carriers of this photosystem are cytochrome b_{559}, pheophytin, and quinines.

Various steps of photochemical reactions, namely light absorption by antenna pigments, migration of light quanta to the reaction center, separation of charge, and movement of electrons, are best explained by **Z scheme**. The scheme explains the participation of PS I and PS I in the synthesis of ATP and formation of NADPH. Water is the ultimate natural electron donor that donates electrons to Q. The process if photosensitized by the oxidized to a strong oxidant capable of splitting water to liberate oxygen through the oxygen-evolving enzyme system. Similarly, P700, the reaction center of PS I on excitation loses electrons to P430 and is oxidized. The reduced Q and the oxidized P700 are reoxidized and reduced, respectively, due to the continuous flow of electrons from water of $NADP^+$ in the presence of NADP reductase. In addition to water, a number of exogenous electron donors have been identified and are known to donate electrons at specific sites of the electron-transport chain. Exogenous electron donors, acceptors, and inhibitors are used extensively to characterize the specific sites of the electron-transport chain of the thylakoid membrane.

Photoelectron transport in thylakoid membranes is coupled to ATP synthesis through coupling factor (CF_1), a well-characterized extrinsic protein complex of the thylakoid membrane (Strotmann and Bickel-Sandkotter, 1984). Thus, light-driven electron transport in thylakoid membranes generates NADPH and ATP, the two essential molecules required for fixation of carbon dioxide to sugar through a series of enzymes in the Calvin cycle. Although most of the enzymes of the Calvin cycle are well studied, rubulose 1,5-bisphosphate carboxylase/oxygenase (Rubisco) has drawn the attention of many workers because its complex structure, regulation of its synthesis, and susceptibility to different environmental stress. The enzyme is also a major protein of green leaves.

DISASSEMBLY OF PS I AND PS II AND LOSS OF ELECTRON-TRANSPORT EFFICIENCY

Senescence causes an extensive disruption of the thylakoid system including the unstacking of grana and disappearance of photosynthetic pigments. This kind of structural degradation could logically lead to a loss in the capacity of chloroplasts for photoinduced electron transport. A decline in the rates of photochemical reactions in chloroplasts undergoing senescence has been demonstrated by many authors using different plant systems (Biswal *et al.,* 1983b; Woolhouse, 1984b). There are, however, different views on the nature of the loss of photochemical reactions. For example, Camp et al. (1982) have shown relative stability in whole chain electron-transport efficiency when it is measured per chloroplast as compared to a leaf area basis at the initial stage of senescence.

The above findings may suggest that a loss in photochemical potential during senescence could be a due to loss of whole chloroplasts rather than the loss of thylakoid organization. On the other hand, the loss in the capacity for photochemical reactions has also been attributed to senescence-induced temporal disorganization of thylakoid ultrastructure, loss of chlorophyll, and physical changes in thylakoid membranes (Choudhury and Biswal, 1979a, 1984a; Hilditch *et al.,* 1986). A gradual decay of membrane structure and consequently a decline in the photochemical potential of chloroplasts are, however, compatible with sequential degradation of thylakoid structure as revealed by electron microscopic studies.

Senescence induced-loss of the photochemical reactions associated with PS I and PS II has been demonstrated to be sequential (Biswal and Mohanty, 1976a; Jenkins *et al.,* 1981; Choudhury and Biswal, 1984a Choe and Whang, 1986; Grover *et al.,* 1986). Senescence of detached barley leaves was shown to cause a relative instability of PS II of thylakoid membranes (Biswal and Mohanty, 1976a). A preferential loss of chlorophyll *b,* a major pigment of PS II, is demonstrated during senescence of mustard cotyledons treated with continuous darkness (Biswal *et al.,* 1983a) suggesting a loss of LHCP and disassembly of PS II of thylakoid membranes (Bennett, 1981; Lichtenthaler *et al.,* 1981). This could be correlated with the loss of PS II photochemical reactions during senescence of leaves (Biswal and Mohanty, 1976a; Choudhury and Biswal, 1984a), algae (Kulandaivelu and Senger, 1976a), and isolated chloroplasts (Choe and Whang, 1986). Severe damage to the oxygen-evolving complex might also be a factor in the loss of the photochemical potential of PS II of thylakoid membranes (Choudhury and Biswal, 1979b; Hilditch *et al.,* 1986).

On the other hand, Sayeed et al. (1985) have demonstrated that during senescence of chloroplasts in detached wheat leaves, a relatively rapid loss of Chlorophyll *a* occurs indicating the disintegration of the PS I chlorophyll-protein complex. This is further confirmed by their observation of the relative instability of PS I photochemical reactions. The preferential loss of PS I is correlated with the senescence-induced disappearance of stroma lamellae before damage to the lamellar system associated with grana thylakoids membranes is attributed to the inactivation or loss of ferredoxin-NADP reductase, FRS, and plastocyanin (Postius, 1971; Eytan *et al.*, 1974) and a loss of the PS I reaction center (Jenkins *et al.*, 1981). A simultaneous decline in the primary photochemical reactions associated with both PS I and PS II of chloroplasts has recently been observed during senescence of a non-yellowing mutant of *Festuca pratensis* (Hilditch *et al.*, 1986).

In addition to the differential loss of the photochemical reactions of PS I and PS II, Biswal and Mohanty (1976a) have proposed specific alterations of the sites in the electron-transport chain between water and the PS II reaction center of senescent chloroplasts. Their observations suggest that diphenylcarbazide (DPC) and Mn^{2+} feed electrons at different sites in the electron-transport chain, DPC donating at a site closer to the PS II reactions center. The site for entry electrons from Mn^{2+} remains further away from the reaction center. During senescence, the site for Mn^{2+} disappears faster than the site for entry of electrons from DPC. Experiments involving the application of different electron donors and acceptors specific to different sites of the electron-transport chain and quantitative measurement of electron-transport components suggest that a major change limiting electron transfer in senescent chloroplasts lies between the two photosystems rather than within them. An inactivation of plastoquinone (PQ), as suggested by Woolhouse and Jenkins (1983), and loss of the cyt *b*-cyt *f* complex, as indicated by David-Ben et al. (1983), may be the factors involved in the loss of electron-transport efficiency.

DAMAGE TO THE OXYGEN-EVOLVING COMPLEX

A decline in oxygen evolution and a restoration of the senescence-induced loss of PS II-mediated DCIP photoreduction with exogenous electron donors are indicative of the limitations imposed by a damaged oxygen-evolving system on the efficiency of the electron-transport system of thylakoid membranes during senescence. Although the precise geometry and the mode of function of the oxygen-evolving complex are not yet clearly known, recent findings propose the participation of a number of polypeptides that constitute the O_2-evolving complex (Murata and Miyao, 1985; Renger and Govindjee, 1985). Some of the polypeptiedes are known to be intrinsic and are embedded in the lipophilic environment of the membrane, and some are extrinsic, obviously vulnerable to externally applied reagents. The model proposed by Raval and Biswal (1985) provides the details of the arrangement of the polypeptides and their association with Mn^{2+} to form an active oxygen-evolving enzyme complex. Some of the polypeptides in this model are shown to be extrinsic.

The orientation of extrinsic proteins proposed in the model is supported by the experimental data of Murata and Miyao (1985) who have shown three extrinsic proteins, namely 33-, 24-, and 18-kDa polypeptides associated with the oxygen-evolving complex. Mild treatments with Tris buffer or high pH are shown to release these polypeptides into the aqueous phase from the membranes (Kuwabara and Murata, 1982; Murata and Miyao, 1985). These experiments suggest that extrinsic proteins that are exposed to the aqueous phase and, not well protected in the membrane milieu are labile and highly susceptible to stress. It is, therefore, logical to suppose that these polypeptides could behave in a manner similar to the soluble stroma proteins in terms of their accessibility to proteolytic enzymes during senescence. The degradation of these extrinsic proteins could load to a decline in the efficiency of the oxygen-evolving complex. Alternatively, senescence-induced loss of Mn^{2+} ions could be a factor in the damage to the oxygen-evolving complex. Although the precise role of Mn^{2+} in the oxygen-evolving complex remains unknown, two different pools of the metal ion in the complex are suggested. One pool of the ion is tightly bound to the intrinsic polypeptides and acts as the change accumulator at the active site of the water-splitting system. The second pool of Mn^{2+}, which is bound to the extrinsic protein, is relatively unstable and seems to be susceptible to different treatments (Raval and Biswal, 1984; Raval *et al.*, 1985). The extrinsic 33-kDa polypeptide has recently been shown to be responsible for stabilizing Mn^{2+} (Murata and Miyao, 1985). Heat treatment causes the release of Mn^{2+} results in the complete inactivation of oxygen evolution (Nash *et al.*, 1985). It is possible that during senescence a loss in Mn^{2+} may result in damage to the oxygen-evolving complex. Supporting this view is the report that senescence causes a quantitative loss of chloroplast manganese (Margulies, 1971).

LOSS OF PHOTOPHOSPHORYLATION

A few reports are available on the senescence-induced loss of the ability of plastid membranes to synthesize ATP (Hernandez-Gil and

Schaedle, 1973; Biswal and Mohanty, 1978; Camp et a., 1982; Choudhury and Biswal, 1984a). There are several possible explanations for the mechanism of senescence-induced loss of photophosphorylation. A decline in the efficiency of the electron-transport system during senescence as reviewed earlier may limit the ATP-synthesizing ability of the membrane. This is also evident from the observation of the senescence-induced loss of activity of the thylakoid-bound Mg^{2+}-ATPase (Choudhury and Biswal, 1984a). The activity of this enzyme reflects the energization status of thylakoid membranes induced by photoelectron transport.

An inactivation or quantitative loss of coupling factor proteins could also lead to a decline in photophosphorylation (Thomas, 1983; Choudhury and Biswal, 1984a). A change in the aggregation pattern and conformation of the coupling factor may result in a decline in its functional efficiency. In fact, Choudhury and Biswal (1984a) have demonstrated a structural modification on the surface of the coupling factor as probed by Atebrin fluorescence during senescence of maize chloroplasts. They have also demonstrated a quantitative loss of coupling factor protein that contributes to the loss of the ability of chloroplasts to synthesize ATP. A quantitative loss of coupling factor protein has also been demonstrated by Camp et al. (1982) during senescence of wheat chloroplasts. They have shown that chloroplasts isolated from mature leaves contain more α and β protein subunits of the coupling factor than chloroplasts from senescent leaves, suggesting a senescence-induced selective loss of coupling factor subunits.

LOSS OF RUBISCO ACTIVITY

The modification of enzymes of the stroma has been examined during chloroplast senescence (Batt and Woolhouse, 1975; Gordon *et al.*, 1978; Makino *et al.*, 1985; Grover *et al.*, 1986). The activities of NADP-dependent glyceraldehydes 3-phosphate dehydrogenase, phosphoglycerate kinase, ribose 5-phosphate isomerase, phosphoribulokinase, and ribulose 1,5-biphosphate carboxylase reportedly decrease during senescence. Ribulose 1,5-biphosphate carboxylase has been shown to be most susceptible to senescence (Lauriere, 1983). There is, however, controversy concerning the relative susceptibility of Rubisco compared to that of chlorophyll and photochemical reactions. For example, it has been suggested that the loss of Rubisco activity occurs earlier and faster than the loss of chlorophyll or photoelectron transport (Biswal *et al.,* 1983b). On the other hand, the enzyme shows more stability than chlorophyll during senescence of the chloroplasts of wheat flag leaves (Hall *et al.,* 1978). Similarly, Wittenbach et al. (1982) have demonstrated that photosynthesis declines faster than Rubisco activity during senescence of chloroplasts isolated from the protoplasts transport and carbon dioxide fixation are also available (Hernandez-Gila and Schaedle, 1973; Gordon *et al.,* 1978). Grover et al. (1986) have suggested a simultaneous triggering of the loss of chlorophyll and Rubisco activity during dark-induced senescence of wheat leaves.

MECHANISM OF LOSS

It is well known that Rubisco consists of large and small subunits differing in molecular weight. The large subunit is encoded by plastid DNA and is synthesized on 70 S ribosomes of the organelle. On the other hand, the small subunit of the enzyme is encoded by the nuclear gene and the protein is synthesized on 80 S ribosomes of the cytoplasm. The synthesis of the small subunit positively controls the synthesis of the large subunit, suggesting a nuclear control over synthesis of chloroplast proteins (Frosch *et al.,* 1976). The enzyme has been studied extensively during senescence of chloroplasts. The reports on the loss of its activity are rather controversial. Several factors could contribute to the loss of enzyme activity.

Loss of enzyme protein could be factor contributing to the decline in enzyme activity expressed on either a total soluble protein or leaf area basis. Wittenbach (1978) has quantitated the protein immunochemically by using a rocket electrophoresis technique and has reported an enzyme loss constituting about 80% of the total soluble protein even at the initial stage of senescence. Interestingly, at this stage of senescence no significant change in enzyme activity per mg of antibody-recognized carboxylase could be obtained. During senescence of rice chloroplasts, Makino et al. (1983, 1984) have shown that no change in the specific activity of Rubisco occurs when expressed on an enzyme protein basis. However, a decline in the activity is obtained when it is calculated on a leaf area basis. Similar results for the senescence-induced loss of enzyme proteins are reported for barley leaves (Peterson and Huffaker, 1975) and wheat leaves (Mae *et al.,* 1984; Grover *et al.,* 1986). The quantitative loss of

Rubisco during senescence has also been examined by immunological techniques during chloroplast senescence in wheat flag leaves (Peoples *et al.,* 1980). At initial stages of senescence, the protein degrades at a rate almost equal to that of the degradation of other soluble proteins. Subsequently, the enzyme protein is rapidly lost, however, without any change in the specific activity of the enzyme as expressed on an enzyme change in the specific activity sites as probed by the protein basis. No change in the number of active sites as probed by [^{14}C] cyanide-binding assay could be demonstrated. These reports suggest a quantitative loss of enzyme protein as a major factor for the loss of enzyme activity. A senescence-induced decrease in the specific activity of the enzyme has also been demonstrated by many authors in different plant systems (Callow, 1974; Hall and Brady, 1977; Hall *et al.,* 1978; Wittenbach *et al.,* 1982).

The decrease in the specific activity of the enzyme as calculated on the basis of enzyme protein and is obviously the consequence of the senescence induced qualitative modification of the enzyme. This could be an alteration in the enzyme such as a change in the number of active sites and/or modification of the sites. For example, the senescence-induced rapid decline in the absolute specific activity of Rubisco has been attributed to a loss of active sites (Wittenbach, 1978); Wittenbach *et al.,* 1982). A change in the kinetic form of the enzyme could also lead to a lowering of enzyme activity during senescence (Hall *et al.,* 1978).

The results suggesting a senescence-induced loss of enzyme protein or a qualitative modification of the enzyme structure leading to a decline in its specific activity are rather conflicting, which prevents a generalization about the behavior of the enzyme during senescence. The data should be carefully analyzed with regard to the experimental conditions, which seem to be very divergent. Various experimental models are used on different plant systems by different authors. It is also difficult to find suitable probes to characterize the fine structural changes at the active site or other qualitative changes of the enzyme surface induced by senescence. Finally, the system of reference used to express enzyme activity should be considered.

OXYGEN EVOLVING COMPLEX

Photophosphorylation

Photosynthetic Changes

Decrease in rate of photosynthesis with the age of a leaf to decline in dry weight of leaves and their eventual death are the symptoms of leaf senescence. The rate of photosynthesis starts declining as soon as the leaf reaches its full size or becomes mature (senescence leads to profound changes in the pigments composition of the cell, thereby affecting photosynthesis rate). There is gradual loss of chlorophyll and carotenoids, which, results in yellowing of leaves and/or cotyledons particularly during the later stages of senescence. The usual explanation given is that carotenoid are destroyed more slowly than chlorophyll during the degradation of the photosynthetic apparatus. Both the chlorophyll and the carotenoids are located specifically within individual protein pigment complexes in the thylakoids membranes. During senescence, these complexes are gradually destroyed by the oxidation and the enzymatic degradation and thus, pigments are lost. It is observed in barley leaves that dark reaction of photosynthesis is more sensitive to degradation during senescence than electron transport and photophosphorylation, especially cyclic photophosphorylation. Apart from this phosphocarboxylase (main protein of chlorophyll) is the most studied enzyme of the dark stages of photosynthesis during senescence and probably it is one of the most quickly degraded soluble proteins. An acid protein is responsible for degradation of RUBP-carbon.

Respiration

When a plant tissue has reached maturity, its respiration rate will either remain roughly constant or decrease slowly as the tissue ages and ultimately senesces. One might except that if senescence is viewed as a general decline in function, the respiration rate would steadily decline. In fact such a decrease does generally occur, but only quite late in the senescence process, and at first there is a sharp increase. The form and timing of this increase is suggestively like the climacteric rise in the respiration of ripening fruit. Hover and Gustafson (1927) first noticed it in wheat seedling. Its incidence and magnitude vary with the material and the circumstances. In the relatively long-lived *Perilla* leaves (attached to the plant) the respiratory rate is almost constant till 56th day, then rises to double its value just before abscission (Colquhoun, Hillman, Crewe and Bowes 1975). In tomato leaf segments, the rise is preceded by a steep fall for the first 6 days (Feller, Soong, Hageman, 1977). On oat leaves, it starts after about 24 hr and reaches 2.5 times the initial rate in about 2 more days; then it decreases rapidly. In general, the respiratory apparatus of senescing tissues remains until late in senescence then declines rapidly (Hardwick, Wood and Woolhouse, 1968; James, 1953). The mitochondria often swell and decrease in number, but changes generally occur late in senescence (Barton, 1966; Butler, 1967). Decrease in respiration occurs only late in senescence, indicates that cellular energy is required for synthesis of degradation enzymes. Wittenbach et al. (1977) found that there is an increase in protease during senescence.

The fact that (a) respiration decreases only late in senescence and (b) anoxia inhibits chlorophyll and protein breakdown (Martin and Thimann, 1972; Wood and Cruickshank, 1944) indicates that cellular energy is required during senescence, presumably for the synthesis of degradative enzymes. Thimann (1980) found that the level of ATP increases during senescence. ATP retards senescence in light by stimulating the biosynthetic process in general. In the dark, ATP accelerated senescence by providing energy for the biosynthesis of only degradative enzyme and of other possible proteins required for senescence.

At light, the main source of ATP is the photophosphorylation. Thimann (1980) suggested that in the dark ATP must be produced during respiration. If this model is correct, the question is: What determines that the biosynthetic processes. The most plausible possibility is that at light genes, which retard senescence, are expressed (e.g., structural genes for protein present in normal functional mesophyll cells). However, in the dark, genes for degradative enzymes and other possible protein implicated in senescence symptoms are expressed. Independently of these genetic controls in relation with the role of respiration in senescence, the question is if respiration rate increases (when present) are a consequence of other senescence processes or are directly affected by the hormonal changes occurring during senescence. In excised leaves of *Rumex*. Goldthwailte (1974) found that gibberellins and cytokinins inhibit respiratory metabolism and senescence in the dark. However, ABA accelerates senescence and increases the rate of respiration. The effects of GA_3 and ABA on respiration are detected as soon as 3 hr after hormone treatment. This suggests that hormone effects on respiratory metabolism are direct. Sucrose, which stimulates respiration, does not however inhibit chlorophyll or protein breakdown. Thus, increased respiration is not sufficient for senescence. Probably respiration in senescence is higher due to the low GA and cytokinins levels relative to the ABA levels. But it seems that although similarly responding to hormones, respiration and other senescence phenomena (as protein and chlorophyll breakdown) are controlled by hormones through different mechanism.

Low RQ values were detected in senescing leaves (Tetley and Thimann, 1975; Veierskov, 1987). This makes clear that lipids rather than carbohydrates were primary source of respiration substrates.

Nutritional Status

The general simultaneity between monocarpic senescence and reproductive processes, led to Molisch, 1938 to postulate that senescence of lagged organs is produced by a competence between these and the reproductive structures for nutrient uptake. At the end the competence would produce a withdraw of nutrients from vegetative to reproductive organs. In relation with the nutrient redistribution associated with plant senescence, the degradation of proteins and the export of nitrogen by senescing leaves is, probably, the most studied process. However, only a small number of quantitative data are known on the nitrogen supply by senescent leaves to reproductive structures in relation with the nitrogen supply by assimilation *de novo.* By comparing the seed yields in nitrogen and in weight in different crop species, Sinclair and De Wit (1975) concluded that in gramineous, and probably also in leguminous, the seed development requires withdraw of nitrogen from leaves to reproductive structures. In forage gramineous, Salette and Lemaire (1981) found a continuous diminution in the percentage of nitrogen in dry matter during the growth that requires a continuous mobilization of nitrogen from old to young organs. When nitrogen is abundant in the soil, almost all of the final nitrogen content of the forage plant is taken up early in the plant development (Salette and Lemaire, 1981).

According to Simpson and Dalling, 1981 field grown cereals stem contributes 40 per cent leaves 50 per cent and the roots 10 per cent to the grain nitrogen, which is transported by way of the xylem and phloem. In soybean, Abu-Shakra et al. (1978) reported that during the first days of flowering there is an increase in nitrogen assimilation (both as N_2 and nitrate) but this assimilation quickly diminished at the latest steps of seeds development. This behaviour seems to be a general phenomenon (Bethlenfalvay and Phillips, 1977; Pate, 1973; Harper and Hageman, 1972).

Abu-Shakra et al. (1978) suggest the potential use of soybeans variants with delayed leaf senescence in improving the yield in nitrogen and weight of the crop. Finally, depodding surgically modified plants at various stages of seed growth suggests that the senescence signal is generated quite late in seed development when most of the dry weight accumulation (nutrient accumulation) is finished (Noodén, Rupp and Derman, 1978).

In addition, the nutrient drain and diversion hypothesis do not seem to hold for spinach and cocklebur (Krizek, McIlrath and Vergara, 1966; Leopold, 1959; Noodén and Leopold, 1978).

It must be clear that although nutrient redistribution may be a function of senescence, it is not necessary the cause of senescence. It is possible that plants have evolved mechanism for the control of senescence by hormones as they have evolved mechanism for the hormonal control of their correlative development, according to their genetic potentialities and to the changing necessities of the plant. For example, it is probably that in a dioecious plant such as *Spinach* the male plant has no nutritional motivation for monocarpic senescence.

However, the space and nutrient requirements of the future plants, make advantageous the senescence and death of both male and female plants. On the other hand, although deflowering or depodding usually retards senescence, this is not always the case. Defloration in cocklebur (Krizek *et al.,* 1966) or defruiting in corn (Allison and Weinmann, 1970) and in pepper (Hall and Brady, 1977), do not retard (in some cases they do accelerate) senescence. It must also be noted that nutritional competence can hardly explain the senescence preceding the fall death of the leaves of deciduous trees. Probably, although with different details in the mechanism, hormones, sometimes together with nutritional factors, control senescence, which developed in all plants according with a genetic programme (Thomas and Stoddart, 1980). In this regard, it must be noted the independence of the genetic determinants for flowering and fruiting in peas with that of apical senescence.

Proebsting et al. (1978) reported the existence of different genetic lines of peas in relation with flowering and apical senescence. One line, G_2, requires long days together with flowering for apical senescence. The long days requirement for apical senescence is determined by the dominant alleles at two genetic loci. Lines I_1, I_2, and I_3 without one or the two dominant alleles at the two loci, show apical senescence independently of the photoperiod.

A line, G, which only differs from G_2 by the presence of a dominant allele in first loci also requires long days for flowering. In short days G_2 can form flowers without apical senescence, which again rule out any nutritional theory for the control of senescence. Line G requires long days for flowering and for senescence, but a genetic analysis of the other related lines, show the independence of at least some genetic determinants for flowering and for apical senescence.

Hormone Levels

According to Osborne and Thompson (1992) the response of any cell to a change in the level of any of the regulatory substance is dictated not by the chemical alone, but first and foremost by the mode of the genetic expression of the cell. However, the latter is controlled by the environment, hormones and developmental stages of the plant; which also determines the way in which, and the extent to which, the signal is perceived and transduced (Bressan and Handa, 1992). The balance among the different growth regulators controls senescence. Change in hormone level influenced by the development stage of leaves and by other structures of the plant. Hormone level changes may be a primary cause or a secondary effect of senescence in leaves.

It is generally agreed that all plant hormone groups influence senescence (Kelley and Davies, 1988; Noodén and Leopold, 1988). In most plants cytokinins are antisenescence compounds that can postpone plant death (Noodén *et al.,* 1989), gibberellins also have similar effects, but in fewer plant species. Auxins influence senescence and abcission in very complex manners, apparently depending on the auxin source, or depending on the age and receptivity of the tissue (Noodén and Leopold, 1988). When applied to the leaf, auxin often retards senescence, but paradoxically when auxin enters the leaf from the stem, it sometimes promotes senescence (Tamas et al. 1981, Noodén and Leopold, 1988).

Two plant hormones have been suggested as candidates for death hormones: ethylene and abscisic acid (Noodén and Leopold, 1988).

Ethylene is intimately involved in many senescence phenomena, especially flower wilting and fruit ripening. However, there is no clear correlation between ethylene and monocarpic plant death (Mattoo and Aharoni in Noodén and Leopold, 1988). Also the mobility of ethylene as a gas does not fit that of a death hormone (Woolhouse, 1983). However, one could imagine that the ethylene precursor, aminocyclopropane carboxylic acid, is transported, and that ethylene is a kind of secondary product within the individual plant tissue.

Abscisic acid was isolated on the basis of its promotion of abcission and dormancy. The compound is much less active on whole plants, and there is little correlation between ABA concentration and plant death (Tames et al. 1981; Kelley and Davies, 1988, Noodén and Leopold, 1988). It is also difficult to reconcile the large fluctuations of ABA in response to drought and turgor changes with the protracted signaling at the stage of seed maturation.

Abscisic acid (ABA): Abscisic acid was originally isolated as an abscission-causing factor. However, it has since become evident that ABA stimulates abscission of organs in only a few species and that the primary hormone causing abscission is ethylene. ABA, on the other hand, is clearly involved in leaf senescence, and through its promotion of senescence it might indirectly increase ethylene formation and stimulate abscission.

The endogenous concentrations of ABA in developing fruits of barley, wheat, soybean, and cherry tomato change in a similar pattern. ABA rises during the rapid growth stage of the fruit or seed, peaks at the point of maximum fresh and dry weight of fruit, and then declines rapidly with senescence of the plant and with ripening or drying of the fruit (Abdel Rahman, Thomas, Doss and Howel, 1975; King, 1976; McWha, 1975; Quebedeaux, Sweetser and Rowell, 1976, Radley, 1976). Developing cotton and pea fruit show a slightly different pattern, with low levels of ABA during rapid fruit growth and a rise during fruit maturation (Davis and Addicott, 1972; Eeuwens and Schwabe, 1975). During the final maturation of cotton, fruit walls senesce (Davis and Addicott, 1972). The magnitude of the increase in fruit may be quite large. Wheat grains have 40 times as much ABA (per gram fresh weight) at maximum fresh weight than in early development, and soybeans have 15 times as much (Raschae, 1976;

Quebedeaux, Sweester and Rowell, 1976) Soybeans, however, contain 30 times more ABA at maximum fresh weight than wheat (McWha, 1975; Quebedeaux, Sweester and Rowell, 1976). Thus, the later stages of fruit development are accompanied or preceded by a rise in ABA levels. ABA (measured by gas-liquid chromatography) rises sharply in detached tobacco leaves, and then rapidly declines (Even-Chen and Itai, 1975) Increases in endogenous ABA controlled by the photoperiod may play an important role in the natural induction of leaf senescence, especially in tree species undergoing autumn leaf senescence (Noodén and Weber, 1978). The rise in ABA levels during senescence might make this hormone an attractive candidate for the role of general senescence inducer, but not all studies support this. ABA does not increase significantly in attached rice leaves during monocarpic senescence, (Oritani and Yoshida, 1973) and young cockelebur leaves contain relatively high ABA concentrations that decrease throughout the life of the leaf (Raschke and Zeevart, 1976). The primary leaves of the common bean show widely varying levels of ABA during senescence but no consistent change (Colquhoun and Hillman, 1975). In attached bean cotyledons, Dumbroff et al. (1977) have attributed the changes in ABA levels to changes in water potential of the senescing tissue, rather than to differences in senescence activity.

Senescent bean leaves contain less free ABA than young leaves. However, the senescent leaves contain more bound ABA, thereby equalizing the total ABA in the young and senescent leaves (Osborne, Jackson and Milborrow, 1972) In analyzing the senescence related changes in ABA levels, it is essential to consider and control environment factors, particularly including water stress, which is now known to increase ABA levels. Leaf levels of ABA increase rapidly in response to water stress seemingly through *de novo* synthesis (Mittelheuser and Van Steveninck, 1972; Wright and Hiron, 1969, Zeevart, 1977).

A eight hour period of water stress in perennial rye (*Loliunm temulentum*) produces a 30-fold rise of ABA in the accelerated senescence in tobacco (Even -Chen and Itai, 1975). The work of Dumbroff et al. (1977) on senescing bean cotyledons shows that the changing ABA levels are better correlated with internal changes in water potential than with senescence. The acceleration of senescence in tobacco, bean, and sunflower leaves by a brief exposure to high temperature (47 °C, 2 min) is accompanied by increased ABA (Itai and

Benzioni, 1974). Nutrient deprivation, particularly nitrogen stress (which is known to accelerate senescence, Milborrow, 1974) also increases ABA in the leaves of sunflower, especially the older leaves (Goldbach, Goldbach, Wagner and Michael, 1975).

Water stress induces ABA in different subcellular compartment (cytoplasm?) from the nonstress ABA (Lovell, and Sagar, 1972). The stress-induced increases in ABA may be large enough to mask the senescence-related changes, or if variable, they could obscure them. Clearly, stress-induced senescence and ABA need to be controlled if other patterns, especially monocarpic senescence, are to be studied.

The exogenous supply of ABA promotes chlorophyll loss in leaf discs collected outdoors from seventeen tree species plus bracken fern (El-Antably, Wareing and Hillman, 1967). ABA also accelerates yellowing of detached leaves in radish, barley, nasturtium and *Rumex pulcher* (Aspinall, Paleg and Addicott, 1967; Back, Bittner and Richmond, 1972; Beevers, 1968; Karanov, and Pogoncheva, 1976; Srivastava, 1968). Different sensitivities to ABA have been found within a species, leaf discs of some cultivars of soybean respond to ABA more rapidly than others (Sloger and Caldwell, 1970). The physiological status of the leaf material is also important; leaves aged by floating on water showed a greater respond to ABA than freshly excised material (Osborne, 1968).

The effect of ABA on attached leaves is generally less pronounced. In many cases, ABA has no effect on attached leaves of the same species for which it promotes the senescence of excised leaves or leaf discs (El-Anatably, Wareing and Hillman, 1967; Sloger and Caldwell, 1970; Smith, Lyon, Addicott and Johnson, 1968). ABA sprays have no effect on leaf senescence of wheat seedling (Wittenbach, 1977).

Translocation, inactivation, poor uptake, and competition from endogenous hormones are all possible reasons for this difference between attached and detached leaves. In fact, attached leaves sprayed once (or even several times) with ABA solutions may still receive less of the hormone than leaves floating on the solutions. Nevertheless, ABA sprays can accelerate senescence of attached leaves including tree of heaven, Brussels sprouts, and soybean (El-Antably, Wareing and Hillman, 1967; Lindoo and Noodén, 1977; Lindoo and Noodén, 1978; Sloger and Caldwell, 1970). ABA hastens fruit maturation in soybeans, but not enough to account for the acceleration of senescence (Lindoo and Noodén, 1976; Lindoo and

Noodén, 1978). ABA does not induce the senescence of depodded soybeans, indicating that ABA alone cannot completely replace the influence of the pod upon senescence (Lindoo and Noodén, 1976; Lindoo and Noodén, 1978).

The function of ABA as a promoter of flower tissue senescence, including such parameters as wilting, color fading of `blueing', and protein loss, has been established (Borohov, Tirosh and Halevy, 1976; Mayak and Dilley, 1976; Mayak and Halevy, 1972). Here, ABA may act through hastening of autocatalytic ethylene production (Mayalc and Halevy, 1972). In some cases, ABA actually exhibits a senescence-retarding effect.

ABA sprays cause persistence of grape inflorescences, which mimics cytokinin sprays and have been termed the cytokinin sparing effects (Mullins and Osborne, 1970). At 10^{-4} M, ABA retards senesces in radish leaf discs, but this retardation may be due to the disruption of the normal senescence process (Colquhoun and Hillmann, 1972).

In addition, ABA may cause short-term inhibition of senescence of cut flowers by closing the stomata, thereby water loss (Halevy, Mayak, Tirosh, Spiegelstein and Kofranek, 1974). Acceleration of RNA and protein loss in senescing tissues by ABA has been reported several times (Beevers, 1968; Colquhoun and Hillman, 1972; Knypl and Mazureyk, 1971; Paranjothy and Wareing, 1971).

ABA elevates protease and nuclease activity through increased synthesis (or the activation of stored enzyme), but not all hydrolases are increased ABA has been found to block the rise of β-1,3-glucan hydrolase in tobacco leaf disks, while this enzyme increases during the senescence of the water controls (Moore and Stone, 1972). Overall, ABA decreases protein synthesis and increases protein degradation, but the relative contributions of these processes and their responses to exogenous ABA are not clear. Not all of the studies on protein synthesis effects are corrected for changes in the specific activity of the precursor pool or decreased precursor uptake over time (Noodén and Leopold, 1978; Sacher, 1967). Many pieces of evidences suggest that ABA may play an important role as an inducer of senescence, including monocarpic senescence; the picture is far from complete.

Zacarias and Reid (1990) compared the effects of ABA and ethylene on the senescence of leaf discs excised from wild-type *Arabidopsis* and an ethylene-insensitive mutant. Chlorophyll loss from untreated leaf discs of the ethylene-insensitive mutant was much slower than that from the wild type, in the light and in the dark. Ethylene accelerated the chlorophyll loss from the wild-type leaf discs, but it had no effect on the yellowing of the ethylene-insensitive mutant discs. In contrast, treatment with ABA stimulated chlorophyll loss in *both* the wild type and the mutant discs, whether they were held in the light or in the dark. These results indicate that the promotion of senescence by ABA is not mediated through its stimulation of ethylene production (Zacarias and Reid 1990).

Ethylene: Unfortunately, little work has been done to examine the involvement of ethylene in whole plant senescence; perhaps that is because it appears not to play a very important part. The senescence in flower petals (fading, rolling, etc.) is accompanied by the increase of production of ethylene (Akamine, 1963; Kende and Hanson, 1976; Kende and Baumgarther, 1974; Nichols, 1966). Production of ethylene in leaves generally rises slowly until the leaves become senescent and abscise. Flowers also synthesize ethylene, especially just before they fate and wither, and in most species this gas causes their senescence and abcission. Many ethylene effects are accompanied by increased synthesis of enzymes, the kind of enzyme depending on the target tissue. When fruit ripening or flower senescence occurs, several kinds of necessary enzymes are produced. Increased ethylene evolution increases the levels of ABA, which may then decrease further production of ethylene (Mayak and Halevy, 1972). In this case, the ABA could serve to slow the rapid rise in endogenous ethylene, which occurs due to its autocatalytic mode of synthesis (Kende and Baumgarther, 1974). Work on morning glory flowers suggests that ethylene may not be the initial trigger of petal senescence, but only an accelator (Kende and Hanson, 1976).

Although ethylene has profound role in fruit ripening, excised organ senescence and flower initiation in some plants (Nickell, 1979), its precise role in whole plant senescence is still elusive. Recently, Morgan et al (1992) have demonstrated that intact leaves of cotton experiences a climatric like storage in ethylene after visible symptom of senescence appears, which consequently signals for abscission. L-serine promotes the senescence of detached oat leaves possibly through stimulation of the formation of two proteases, which normally increase throughout the senescence of the leaf (Martin and Thimann, 1972). L-serine has only a slight effect on the senescence of isolated chloroplasts, the protease synthesis is assumed to occur on cytoplasmic polysomes. L-cysteine, L-glycine, and L-alanine also promote senescence and protease activity, but they act only at higher concentrations than serine (Martin and Thimann, 1972). The role of endogenous serine in control of senescence remains to be clarified.

Ethylene has well characterized effects on the abscission process, stimulating the synthesis and secretion of cellulases and polygalacturonases (Ables, 1973; Noodén and Leopold, 1978). Ethylene (7 ppm) has no effect on leucine incorporation in *Rhoeo* leaf disks, but there may be a slight inhibition of uridine incorporation into RNA (Sacher and Salminen, 1969).

In morning glory corollas, ethylene accelerates both fading and the rise in RNAse, though the fading comes first (Kende and Baumgartner, 1974). Treatment of cut carnation and *Chrysanthemum* flowers with ethylene, 3 ppm, caused the acceleration of senescence associated changes in protein electrophoretic patterns; the proteins become a more diverse mixture with lower molecular weights than those of fresh tissue (Parups 1971).

Ethylene plays a very prominent role in the senescence of certain plant parts, particularly fruit and petals, and in the abscission process. Ethylene is generally implicated as an inducer in the senescence of flower tissue, which responds to lower concentrations than leaves do. Ethylene at 1 ppm induces chlorosis in rose leaves (Zimmerman, Hitchcock and Crocker, 1931), but up to 25 ppm it has no effect on the detached leaves of *Nymphoides* (Goldney and van Steveninck, 1972).

Etherel (which breaks down into ethylene) sprays have no effect on foliar senescence in wheat seedlings (Wittenbach, 1977). Exogenous ethylene (applied as Etherel or the gas) increases endogenous ethylene and ABA production and overcomes the senescence retarding effects of exogenous cytokinins in flowers (Burg, 1962; Kende and Baumgartner, 1974; Kende and Hanson, 1976; Mayak and Halevy, 1972; Smith *et al.,* 1966).

Recently Morgan et al. (1992) have demonstrated a climacteric like rise of ethylene in intact leaves of cotton after the appearance of visible symptom of senescence, which consequently signals for abscission. L-serine, L-cysteine, L-glycine, L-alanine are also known as senescence promotors.

Cytokinin: Cytokinin activity in the leaves decreases during senescence and may parallel chlorophyll loss (Lindoo and Noodén, 1976; Lindoo and Noodén, 1978; Oritani and Yoshida, 1973; Van Staden, 1973).

Cultivars of rice and soybean with early senescence show correspondingly earlier decreases in foliar cytokinin (Oritani and Yoshida, 1973). While free zeatin declines in rice and soybean leaves, the zeatin derivatives dropped less rapidly, or even rose (Oritani and Yoshida, 1973).

Similarly, senescing *Streptocarpus* leaves show a decline in butanol-extractable cytokinins (free zeatin and nucleosides), while the residual water-soluble cytokinins (Zeatin nucleotides) increase (Van Staden, 1973). Treating isolated leaves of many species with cytokinins will delay their senescence. Although applied cytokinins do not prevent senescence completely, their effects can be dramatic, particularly when the cytokinin is sprayed directly on the intact plant (Van Staden et al. 1988).

Probably, root apices are major sites of cytokinin synthesis and from root they are transported to aerial parts of plant. As the formation of roots haltes senescence in detached leaves starts. Roots produce some senescing-retarding factors. Later, Kulaeva (1962) demonstrated that root exudates of substances retard senescence.

Cytokinin influence the movement of nutrients into leaves from other parts of the plant, a phenomenon known as cytokinin-induced nutrient mobilization. This process is revealed when nutrients (sugar, amino acids etc.) radiolabeled with ^{14}C or ^{3}H are fed to plants after one leaf or part of a leaf is treated with a cytokinin. Later the whole plant is subjected to autoradiography to reveal the pattern of movement and the sites at which the labeled nutrients accumulate. Experiments of this nature have demonstrated that nutrients are preferentially transported to, and accumulate in, the cytokinin- treated tissues. It has been postulated that the hormone causes nutrient mobilization by creating a new source-sink relationship.

Leshem (1984) established a causative relationship among cytokinin, lipoxygenation, anti-oxidation and lipid-associated free radical scavenging. During both normal and induced senescence, there is a rise in lipoxygenase which catalyzes the oxidation of polyunsaturated fatty acids. These include linoleic and linolenic acids, which are of common occurrence in plants and inter alia may be situated in membrane phospholipids. Lipoxygention causes the production of free redicals such as the superoxide, fatty acid and peroxy species. Cytokinin lowers lipoxygenase and superoxide dismutase activities significantly in senescing foliage and is mimicked by the endogenous lipid antioxidant, tocopherol, in its chlorophyll-retaining and lipoxygenase – lowering effects. He indicated that cytokinin interaction with free radicals may occur in following two ways:

1. Cytokinin may act as a direct free radical scavenger.
2. Cytokinin may serve as an incipient preventative of free radical formation by inhibiting oxidation of plant purine compounds which at certain stages of breakdown release superoxide and hydroxyl free redicals.

These observations collectively indicate that prevention of free radicals formation and (or) their direct scavenging should be included in the multifactorial antisenescence mode of action of cytokinin.

To test the role of cytokinin in regulation the onset of leaf senescence, Richard Amasino and his colleagues at the University of Wisconsin transformed tobacco plants with a chimeric gene in which the promoter of the senescence-specific *Arabidopsis* cysteine protease was used to drive the expression of the *ipt* gene (Gan and Amasino, 1995). Since the *ipt* coding sequence in this chimeric gene was transcripted only during leaf senescence, the transformed plants has wild-type levels of cytokinin and developed normally, up to the onset of leaf senescence. As the leaves aged, however, the cysteine protease promoter was activated, triggering the expression of the *ipt* gene within leaf cells just as senescence would have been initiated. The resulting elevated cytokinin levels not only blocked senescence, but also limited further expression of *ipt* gene, preventing cytokinin overproduction. These results strongly suggest that cytokinin is a natural regulator of leaf senescence.

Foliar applications of cytokinin have now been shown to retard senescence in attached leaves of several species including common bean (Adedipe, Hunt and Fletcher, 1971; Fletcher, 1969), oat (Thimann, Tetley and Thanh, 1974), soybean (Lindoo and Noodén, 1978; Schreyer and Noodén, 1975), and wheat (Wittenbach, 1977). Spraying with the auxin NAA delays senescence of attached soybean leaves (James, Anderson and Greer, 1965; Kahanak, Okatan, Rupp and Noodén, 1978). The roots are the main sources of cytokinins for the leaves, a decrease in the level of cytokinin during leaf senescence must be due:

1. To an accelerated cytokinins metabolism in the leaves before their senescence.
2. To an accelerated transport from leaves to other structures.
3. To a decrease in the supply of cytokinins by the roots.

Of these three possibilities the last is the most studied. However no definitive conclusions are still available. In tomato, cytokinin decreases in root exudates early during flower development, but then increases when the flower open (Davey and Van Staden, 1976).

In tomato and tobacco, disbudding (which retards chlorophyll loss from leaves) increases levels in root exudates (Colbert and Beever, 1981). In *Perilla* (Beevers and Woolhouse, 1975) and in sunflower (Sitton *et al.,* 1967) cytokinins increased in root exudates during flowering. One can think in a variant of the nutrients competition theory, reproductive and vegetative structures compete for cytokinins from roots.

The cytokinins content, in turn, determinates if a structure will be a sink or a source. However, the fate of root-cytokinin during the reproductive phase is not clear. At early fruit development, its cytokinins content is high, and then decreases (Noodén, 1980a,b). In soybean fruit development no correlation was found between fruit cytokinin decrease and leaf senescence (Lindoo and Noodén, 1978). Noodén and Lindoo (1978) concluded that root-cytokinin reduction is not the primary causal factor in monocarpic senescence in soybean. Kao (1980a,b) concluded that in soybean seedling, leaf senescence is caused by a competition among leaves for root cytokinin. While endogenous level of cytokinin is reduced in intact older leaves prior to senescence (Singh *et al.,* 1992a,b), exogenous application of the hormone is also effective in the deferment of the senescence of foliar organs (Biswas and Choudhuri, 1980; Noodén, 1980; Biswas and Mandal, 1986 a).

If the etiolated leaves are treated with cytokinin before being illuminated, they form chloroplasts with more extensive grana, and chlorophyll and photosynthetic enzymes are synthesized at a greater rate upon illumination (Lew and Tsuji, 1982). These results suggest that cytokinins participate in regulation of the synthesis of photosynthetic pigments and proteins, along with other factors such as light, nutrition and development.

Auxin: Auxin levels decrease with age in leaves of several species (Shoji *et al.,* 1951; Sweetser and Swartzfager, 1978). Auxin activity drops to low levels in the senescing first-trifoliate leaves of the common bean (Shoji, Addicott, Swets, 1951) but an increase is reported in senescing primary leaves of the same species (Wheeler, 1968). Measurements of auxin by modern physical methods show that the IAA content in leaves of cotton and common bean decreased with age (Sweetser and Swartzfager, 1978), thereby supporting the former on bean leaves.

Hartung (1977) found α-naphthalene-acetic acid to retard senescence of leaf discs and bean pods. Osborne and Halevy et al. (1964) reported that 2,4-D butyl ester (a synthetic auxin largely used as a herbicide) retarded the protoolysis of detached leaves of *Prunus serrulata* (one of the flowering cherries). Auxins have not, however, been generally found very potent in retarding leaf senescence.

In oat leaf, the auxin (IAA was used) retarded senescence only at concentrations around 500 times that of kinetin (Hewett and Wareing, 1973). In wheat seedlings naphthalene-acetic acid has a small effect, but only at 100 mg/L, i.e., over 100 times the effective concentration of BAP, and one which would soon be toxic (Klambt, 1968). Dole et al. (1959) reported a clear auxin effect on tomato leaflets.

A complication is that auxin, at levels only slightly above the physiological, stimulates the production of ethylene, which in some tissues promotes senescence; however, its influence in leaves is not strong, and often indeed it seems to show no effect at all. Thus the influence of auxin, though slight, is probably to be ascribed to the auxin it self.

Auxins can retard senescence of detached leaves in some plants (Noodén and Leopold, 1978; Osborne and Halevy, 1960; Osborne and Halevy, 1964; Sacher, 1959). Auxin treatments do not work with all species in same menner (Noodén and Leopold, 1978; Richmond and Lang, 1957). Therefore, auxin is not strong as cytokinin but it has retarding effects on senescence in some assay system (Sacher, 1957).

Chloroindole auxins: 4-chloroindole acetic acid (ClIAA) derivatives (Engvild, 1987; Katayama et al. 1988) may be death hormone in *Pisum, Vicia, Lens* and *Lathyrus,* because of:

1. ClIAA induces very strong, almost irreversible ethylene production (Ahmad et al. 1987).
2. ClIAA induces death of the apical meristem in pea cuttings (Ahmad et al. 1987).

Five different chloroindole auxin derivatives are present in large amounts in maturing pea seeds, and reach concentrations of several mg kg^{-1} for methyl 4- chloroindole acetate (Engvild, 1987; Katayama et al. 1988).

The five naturally occurring chloroindoles known are 4-chloroindole acid, its methyl ester, the methyl and ethyl esters of malonyl-4-chlorotryptophan, and monomethyl ester of 4-chloroindoleacetyl aspartic acid (Engvild, 1987; Katayama et al. 1988). There are of course problems with chloroindoles as death hormones or endogenous herbicides. One problem is the postulation to two distinct auxin groups with opposite physiological properties: growth inducers and endogenous herbicides. Another problem is that chloroindole auxins have so far only been found in pea and related species, but not in soybean or kidney bean (Engvild, 1957). Despite of some attempts, none of these have been identified yet; and they may be death hormones.

Morris et al. (1987) found a chlorine containing compound with both auxin and death hormones like characteristics in the slime mold *Dictystelium discoideum.* The chlorinated phenolic compound was first named DIF-1 (differentiation inducing factor); together with cyclic AMP it induces stalk cell morphogenesis, including cellulose wall formation, cell elongation, vacuolization, and eventually the stalk cell dies. At present, it is not clear if the compound induces cell death as such or just launches the cell on and irreversible path.

Gibberellin: Parallel decrease in content of gibberellins in senescence has been reported in *Tropaeolum* leaves (Chin and Beevers, 1970). A similar decrease occurs in lettuce leaves under water stress. Endogenous gibberellin-like components decline during the senescence of detached *Nastursium* and *Dandelion* leaves.

The amino acid, arginine reverses the senescence promoting effect of serine in oat leaves (Shiboaka and Thimann, 1970) but the effective concentration is probably too high for an *in vivo* significance of the experiments. The retention of chlorophyll in barley leaf discs is enhanced by the polyamines - putrescine, spermidine and spermine.

In leaves of *Taraxacum* (Fletcher and Osborne, 1966) and in fronds of *Lemna trisulca* (Bata and Nescovic, 1974) gibberellic acid powerfully retards senescence, but kinetin has no or only a low effect on senescence. Gibberellic acid was demonstrated to delay development of autumn coloration when applied to several species of deciduous trees (Brian, Petty and Richmond, 1959). Subsequently, gibberellins were shown to delay senescence of detached leaves and leaf segments, but not all species respond to gibberellins (Beevers, 1966; El-Antably, Wareing and Hillman, 1967; Fletcher and Osborne, 1965; Noodén and Leopold, 1978; Whyte and Luckwill, 1966).

Therefore, it seems clear those cytokinins are the most frequently found retardants of senescence in leaves. Gibberellins also retard senescence in some plant. The role of auxin is not so clear probably auxin can sometimes interacts with other retardants in retarding senescence.

Some recent results also support the above hormone receptor theory (Nissen, 1988; Jones, 1990; Mishra, 1992). Saunders (1990) has presented a four-component model of phytohormone action. These are as follows:

1. Perception of stimulus by a receptor
2. Activation of an intercellular second messenger
3. Cascade of amplification of the message
4. Physiological response

Nevertheless, the progress of hormone research on the above lines is far less rewarding, as yet. In addition to the isolation of some specific hormone binding protein (Hall et al 1984; Sisler, 1984; Goodwin and Mercer, 1986; Romanov *et al.*, 1990) some second messengers in hormone action have also been isolated (Roberts and Hooley, 1988).

The role of Ca-calmodulin as second messenger for the regulation of senescence has been attracting interest of some plant physiologists (Leshem *et al.*, 1985; Poovaiah, 1988). Cytosolic calcium hastens the process of senescence through calmodulin: Phospholipase-A_2 complex (Leshem *et al.,* 1985). Intrusion of Ca into cytosol appears to play a pivotal regulatory mechanism of plant catabolism because calmodulin increases free fatty acid and ethylene and both are involved in senescence development (Halevy and Whitehead, 1989; Whitehead and Halevy, 1989). Interestingly, ABA also plays role in intrusion of Ca into the cytosol (Roberts and Hooley, 1988). Some recent reports indicate that transporting of Ca into the cell also activates ABA biosynthesis through signal-transduction chain (Owen, 1988; Skriver and Mundy, 1990). This might be attributed to an autocatalytic effect of ABA.

Paradixically, flower induction is inhibited by a reduction in cytosolic calcium (Friedman *et al.,* 1989). Thus, induction of flowering and senescence might be mediated through intrusion of Ca^{2+} into the cytosol.

Control of GA and Auxin on apical senescence of pea

Pea (*Pisum sativum* L.) lines G_2 (dwarf) and NGB1769 (tall) (Sn Hr) produce flowers and fruit under long (LD) or short days (SD), but senesce only under LD. Endogenous gibberellin levels were inversely correlated with photoperiod (over 9-18 h) and senescence: GA_{20} was 3-fold and GA_1 was 10-11 fold higher in flowering SD G_2 shoots, and the vegetative tissues within the SD apical bud contained 4-fold higher levels of GA_{20}, as compared with the LD tissues. Prefloral G_2 plants under both photoperiod has GA_1 AND GA_{20} levels similar to the flowering plants under LD. Levels of IAA were similar in G_2 shoots in LD or SD; SD apical bud vegetative tissues had a slightly higher IAA content. Young floral buds from LD plant had twice as much IAA as under SD. In NGB1769 shoots GA_1 decreased after flower initiation only under LD, which correlated with the decreased growth potential. It is suggested that higher GA_1 content of G_2 and NGB1769 plants under SD conditions is responsible for the extended vegetative growth and continued meristematic activity in the shoot apex. This and the increased IAA level of LD floral buds may play a role in the regulation of nutrient partitioning, since more photosynthate partitions to reproductive tissue under LD condition, and the rate of reproductive development in LD peas is faster than under SD.

Under LD conditions fruit development in G2 pea plants results in apical senescence, but under SD conditions apical growth continues, despite the presence of a large number of fruit. This is dependent alleles at the Sn and Hr genetic loci. The photoperiodic response of G2 plants is nonphotosynthetic and quantitative, such that a brief exposure of light during the night does not reverse its SD growth habit (Proebsting *et al.,* 1976). The growth rate of developing seeds on G2 plants is more rapid under LD than under SD conditions, which is indicative of a greater nutrient drain exerted on LD plants by their reproductive structures. However, when some of the flowers on LD-grown plants were removed so that the total increase in seed weight on LD-grown plants was the same as or even lower than SD-grown plants, LD G2 plants still senesced, whereas plants under SD conditions did not (Gianfagna and Davis, 1981). Thus, nutrient demand of seeds alone cannot account for the induction of senescence.

Exogenous GAs can delay apical senescence of pea plants (Davis *et al.,* 1977; Proebsting *et al.,* 1978). Among applied plant hormones, including BA, NAA, GA_3, GA_{20}, and GA_9 only GA_3 delayed the senescence of LD-grown G2 peas indefinitely, whereas GA_{20} had a moderate effect. Plants treated with GA_3 produced a full complement of fruits. They were nearly senescent, however, the plants resumed active growth.

Application of GA_3 also caused extensive flower abortion as well as increased stem length. Proebsting et al. (1978), using the lettuce hypocotyl bioassay. They found that SD-grown G2 pea shoots contained a relatively large amount of a polar GA (GA_E) when compared with those of LD-grown G2 plants. The appearance of this GA_E was correlated with the continued, vigorous growth of SD-grown G2 peas. In leaves of photoperiod-insensitive lines (I types), short photoperiods increased neither the GA_E content nor the growth potential to any appreciable extent. Since this GA_E co-chromatographed with authentic [3H]GA_1 and was biologically active, they suggested that GA_E might be GA_1 and that the presence of an increased amount of this GA was responsible for the extended growth potential of SD-grown G2 pea. Lockhart and Gottschall (1961) also noted the fact that GA_3 delayed apical senescence in Alaska peas, but they concluded that the effect was indirect. Ecklund and Moore (1968) confirmed the senescence-delaying effect of GA_3 on the apical bud of pea plants and were able to show that AMO-1618, an inhibitor of GA biosynthesis, accelerated senescence in Alaska peas.

The involvement of IAA in the processes of monocarpic senescence has received less attention. Nonetheless, Nooden et al. (1979) found that foliar applications of NAA, together with BA, prevented seed-induced monocarpic senescence in soybeans. The auxin content of soybean leaves was found to decline with leaf age and was higher in depodded soybean plants, compared with intact ones (Hein *et al.,* 1984). Woolhouse (1983) proposed that the auxin content of the plant decreases following the conversion of all of the vegetative apices to inflorescences, leading to a reduction of cytokinin synthesis in the root.

The maintenance of higher concentration of GA_{20} and GA_1 in the shoots and of GA_{20} in the vegetative tissues of the apical bud correlated with the SD-induced prevention of senescence in G2 and NGB1769 peas. Prior to LD-induced senescence there was a decline of GAs in the vegetative tissue, combined with an increase in IAA in the more rapidly growing flower buds. It was suggest that these hormonal changes directly or indirectly regulate the continued growth or senescence of the apical bud of the pea plant.

In the vegetative tissues within the apical bud it was noted an inverse correlation between GA_{20} and apical senescence. Even though GA_1 is regarded as the prime GA growth regulator, Ross et al. (1992), noted a correlation between the level of GA_{20} and internode length within a genotype, although not across genotypes. This is most likely because the level of GA_1 is in part regulated by the activity of GA 20-oxidase (Heden and Croker, 1992), which catalyzes three reactions that lead to the production of GA_{20}. Thus, the level of GA1 may be linked to the level of its precursor, GA_{20}, when the level of GA_{20}- 3β-hydroxylase remains constant. The level of GA_{20} in apical tissues of G2, therefore, probably reflects the level of GA_1 in these tissues.

The delay of apical senescence of G2 peas has been shown to be correlated with the presence of a graft-transmissible factor (Davis *et al.,* 1977; Proebsting *et al.,* 1977). Proebsting et al. (1977) suggested that this factor, produced in large quantities only under SD conditions and in the presence of both Sn and Hr loci, counteracted the effects of the senescence stimulus. Murfet and Reid (Mufet, 1971, 1973; Murfet and Reid, 1973) have also presented evidence that the Sn and Hr alleles result in the production of a hormone under SD conditions that favors vegetative growth in peas. Application to GA_3 or GA_1 to LD-grown G2 peas results in active vegetative growth as well as extensive flower abortion. Under SD, application of microgram quantities of GA_3 or GA_1 to G2 peas sporadically causes a reversion to the vegetative state and subsequent abortion of all of the flowers.

Proebsting et al. (1978) proposed that the prevention of senescence in G2 peas under SD growth conditions occurred because the level of a biologically active, polar GA (GA_E), regulated by the presence of the dominant alleles Sn and Hr, is sufficiently high to overcome the senescence stimulus that is associated with flower and fruit development. The hormone regulates the prevention of apical senescence in both G2 and NGB1769 peas and is probably GA_1, as was previously suggested by Proebsting et al. (1978). This is based on several observations.

First, the decrease in the levels of GA_1 under LD preceded the onset of apical senescence, whereas the higher GA_1 level under SD conditions is associated with the enhanced growth potential of the apical bud in both lines of peas. For instance, under a 9-h photoperiod G2 pea shoots contained the highest amount of GA_1 (3.4 ng/g fresh weight) and grew very vigorously. The growth potential decreased in parallel with the levels of GA_1 as the latter decreased with an increased photoperiod.

Second, there is less GA_1 in LD G2 than in LD NGB1769, and senescence is more rapid in G2 than in NGB1769. By contrast, GA_1 in nonsenescing, postfloral SD G2 is higher than in postfloral LD NGB1769 in which internode length declines as development proceeds toward senescence, even though NGB1769 is genetically tall (Le).

Third, among the plant hormones tested so far, only GA_3 and GA_1 have the ability to prevent the senescence of LD-grown peas.

It is interesting that not only the appearance of increased endogenous GA_1 levels correlates well with the prolonged or nonsenescing growth response seen under SD conditions, but also that the time at which GA_1 reaches a certain low level under LD following the start of reproduction may account for the time of apical bud death in different lines of peas. For instance, NGB1769 has a high GA_1 content (8 mg/g fresh weight) during the vegetative growth stage and this level goes down progressively only under LD conditions after the initiation of reproductive growth. It usually takes 3 or 4 months for the level of GA_1 to reach 0.3 to 0.4 mg/g fresh weight, and the growth of the apical bud will come to a full arrest in approximately another month. In the case of G2 peas, the level of GA1 is 0.2 to 0.5 mg/g fresh weight during the vegetative growth stage and under LD conditions these peas will grow for only 1 month or 40 d after the start of flowering before the senescence of the apex.

The delay of senescence in G2 peas correlates with a slower reproductive development under SD conditions (Kelly and Davies, 1986). Kelly and Davies (1988a) found that a higher percentage of leaf-exported carbon was transported to the apical bud and the leaf primordia within the apical bud and a lower percentage was transported to the young fruits and the flower buds within the apical bud under SD compared with LD conditions. They concluded that the relative strength of the young reproductive and vegetative sinks is strength (Gifford and Evans, 1981; Kelly and Davies, 1988b; Sklensky and Davies, 1993).

The level or application of IAA and GA has been previously correlated with sink strength in other systems (Patrick, 1979; Patrick *et al.*, 1979). One possible mechanism by which this may occur is by increased hydrolysis. For example, in etiolated pea epicotyls GA application enhanced growth, invertase activity, and sugar accumulation (Miyamoto *et al.*, 1993), whereas IAA has been shown to have similar effects in other situations (Morris and Arthur, 1984, 1986). Therefore, it was proposed that the levels of these hormones also regulate photosynthate partitioning within the apical bud of peas. The decreased IAA level in SD G2 floral buds compared with that of LD floral buds may be viewed as an indication of weakened sink strength and could, therefore account for the slower reproductive development seen in SD G2 peas. The increased GA, content in SD G2 shoots on the other hand, may be responsible for maintaining the active growth of the vegetative apical bud relative to the floral buds. Under LD conditions the increase in IAA levels in the flower buds, combined with the decrease in GA levels in the vegetative tissue, would lead to an increased partitioning of nutrients to the reproductive structures and enhance the senescence of the apical buds. Fruits, or a senescence factor from them, are not central to senescence induction (Hamilton and Davies, 1988a, 1988b; Kelly and Davies, 1988b). Their ability to cause nutrient depletion of the apex is determined by their sink strength relative to that of the apex. Reid (1980) also suggested that senescence in peas was induced early in the reproductive phase and was only enhanced by the fruit.

Internode length in peas has been correlated with the content of GA_1 in the expanding internode (Rose *et al.*, 1992; Reid and Howell, 1995). The internode length of SD-grown G2 peas did not increase relative to the LD-grown plants, despite the fact that GA_1 content increased more than 10 times under SD conditions. This may de due to a reduced response of SD G2 to GA_1, since the overall GA_1, content in SD may not exceed a threshold level that is required to induce further stem elongation. Exogenous GA_1 or GA_3 did produce a similar amount of elongation growth in G2 shoots under SD or LD conditions, but applications were at the microgram level, more than 100 times the amount present in the growing part of a SD shoot.

The capacity to respond by increases in length may also decline with age. Using a normally senescing, tall line of peas, Roes et al. (1992) noted that even though exogenous GA_3 caused proportionally more elongation when growth had almost stopped prior to apical arrest compared with when the plants were growing at a maximal rate, GA_3 did not have the ability to restore internode length to anywhere near the values seen earlier in development.

On the other hand, NGB1769, the tall genotype, displayed a positive correlation between the internode length and the GA_1 content during ontogeny, which is very similar to that noted in the different genotypes by Ross et al. (1992). There are other possible reasons why length. The elevated GA_1 may have been localized in cells or tissues associated with developing primordia and young leaves but not in the expanding internodes, or the increase in stem elongation may involve IAA in addition to GA_1.

Law and Davies (1990) reported a good correlation between IAA level and the stem length of several different pea genotypes, including slender lines in which GA is lacking. Yang et al. (1993, 1995) demonstrated that applications of IAA enhance stem growth in intact pea plants, with the effect being more pronounced in dwarf plants containing lower auxin levels (Law and Davies, 1990). Therefore, it is clear that both GA and IAA are needed in stem elongation, so that when one of them is missing or present at low level the elongation response will be diminished. In the case of G2 peas, both GA_1 and IAA levels were low during the vegetative growth stage and displayed the dwarf phenotype. After flower initiation, the amount of GA_1 in SD-grown pea shoots increased, but the level IAA remained similar in LD and SD. Higher levels of exogenous GA_3 or GA_1 promoted stem elongation as well as IAA accumulation. The IAA level was several fold higher 3 d after GA application.

The senescence of the apical bud in peas, which is the first symptom of the coming demise of the whole plant, is preceded by a decrease of GA_1 in the shoot and GA_{20} in the vegetative tissues of the apical bud, and an increase of the IAA level in the young flower buds. This correlates with a shift in the partitioning of photosynthate from the vegetative to the developing reproductive structures within the apical bud. It was suggest that short photoperiods result in the increased biosynthesis of 13-hydroxylated GAs in the vegetative tissues within the apical bud and that the resulting higher levels of GA_1 enhance the sink activity of these tissues such that vigorous apical growth continues. Flower buds develop under these conditions but much more slowly than under senescence-promoting conditions. Under senescence-promoting (LD) conditions fewer 13-hydroxylated GAs are produced, and vegetative vigor declines at the same time that the IAA level of the buds increases. This leads to more photosynthate moving to the reproductive tissue so that the flower buds grow rapidly at the expense of further vegetative growth, leading to the senescence of the apical bud. Senescence is a consequence of nutrient diversion to the young, developing reproductive structures. Growth of seed enhances the process but is not the cause of the process.

Jasmonic acid: Jasmonic acid derivatives may be death hormones have following effects:

1. Jasmonic acid methyl ester is a senescence inducer in some assays it is stronger that ABA (Sembdner and Klose, 1985).
2. Jasmonic acid widely distributed in plants (Sembdner and Klose, 1985).
3. Soybeans contain high concentration in vascular bundles of the pericarp (Lopez et al. 1987).
4. Jasmonic acid effects can often be alleviated or reversed by cytokinins. This is in agreement with effects of cytokinins on monocarpic senescence (Sembdner and Klose, 1985).
5. Jasmonic acid methyl ester induces ethylene formation (Saniewski et al. 1987)

Jasmonic acid and its methyl ester have been found to occur naturally in a wide range of higher plants and are believed to be derived from fatty acids. Methyl-jasmonate has been shown to promote the senescence of excised oat leaves. Treatment of barley leaf segments with jasmonate or methyl-jasmonate causes not only the acceleration of senescence but also the synthesis de novo of a number of abundant proteins (Weidhase *et al.,* 1987). Muller-Uri et al. (1988) proposed that the protein synthesis induced by methyl-jasmonate is a cause rather than a consequence of senescence. The jasmonate induced proteins (JIP) have been identified. In soybean, two of the induced proteins are vegetative storage proteins (Anderson *et al.,* 1989) suggesting that jasmonate maybe involved in controlling the partitioning of leaf nitrogen into storage form. Chen et al. (1998) have shown that the effect of methyl–jasmonate on rice leaf senescence is associated with ammonium accumulation. It is speculated that methyl-jasmonate might act through the regulation of endogenous ABA level. Wang and Kao (1999) have suggested that endogenous ABA levels are not linked to methyl – jasmonate induced senescence of detacted rice leaves. But the current view is that jasmonates are the part of the stress signal transduction pathway rather then being a direct cause of senescence.

Brassinosteroids: In 1979 a novel plant growth-promoting steroidal lactone termed brassinolide was found in the pollen of rape (*Brassica napus*). Since then chemists and biologists have isolated brassinolide and identified its related compounds in various higher and lower plants. As a result of extensive investigations, brassinosteroids were found to show characteristic physiological actions on the growth of plants in micro-quantities. Therefore, brassinosteroids can be regarded as a new class of plant hormones in addition to auxin, GA, cytokinin, ABA, ethylene and jasmonic acid with an independent role in the growth regulation of plants. Recently, it has been reported that Arabidopsis mutants deficient in brassinosteroid biosynthesis exhibit delayed leaf senescence (Clouse, 1996).

Free Radicals and Ions: Some of the inorganic ions reportedly regulate chloroplast senescence (Poovaiah and Leopold, 1973; Rao and Swamy, 1977; Thompson *et al.,* 1982; Raval *et al.,* 1982,1984; Thimann, 1985a; Panda *et al.,* 1986; Biswal and Choudhury, 1987). The senescence-delaying action of Mg^{2+} is explained in terms of its stabilizing effect on the stacking of thylakoids (Raval and Biswal, 1984), which are known to unstuck during senescence cations such Mg^{2+} and Mn^{2+} have been shown to regulate the level of thylakoid lipid peroxidation and modify the process of chloroplast senescence

(Panda *et al.,* 1987). Liberman and Wang (1982) have suggested that Ca^{2+} and Mg^{2+} interact with the surface of plasma membranes, rigidify them, and consequently delay the process of cellular senescence. The action of Ca^{2+} however, is not comparable with that of Mg^{2+} because of the specific affinity of the former to calmodulin. Calcium ion is also known to stimulate the process of senescence when it enters inside the cell (Leshem *et al.,* 1984)

Free radical reactions are undoubtedly involved in ageing, age-related disorders and death of the organs or organisms and there are also in built enzymatic and non-enzymatic defence mechanisms for protection against free radicals, resulting in senescence. This theory excludes all other reactions, not connected with free radicals, resulting in senescence. Like all other developmental processes, it is still not very clear that how the cascade of events at the molecular level is set in motion leading to the accumulation of free radicals. Nevertheless, this is true that if the free radicals are contained, the longevity of the organism can be prolonged.

Free radical theory of ageing has attracted considerable attention of scientist. As suggested by a number of workers, ageing in plants is accompanied by a progressive shift from a reduced to an oxidative state in tissues. In the last decade and in recent years, free radicals have been equally implicated senescence of leaves and there is now increasing tendency to view the effect of phytohormones on senescence as mediated through modulation of metabolism of free radicals and also their subsequent deteriorating action on membranes caused by lipid peroxidation. As stated, the onset of senescence marks the end point in the transition towards on oxidative state in tissues with the diminution of sulphydryl gradient and oxygen utilization could accelerate the transition of the tissue towards anoxidative state and thus favour the development of senescence. So many activated oxygen species, such as singlet oxygen ($^1O_2$), superoxide (O^{-2}) hydroxylradicals (OH) and hydrogen peroxide (H_2O_2), are known to accumulate in plant tissues during the senescence process. The free radicals may be generated either by xenobiotic (through environmental pollution) or biotic (leakage from sites of irregular cellular metabolism) means (Leshem, 1984).

The molecular oxygen species (O_2) is not involved in the oxidative processes in ageing tissues, since metabolites commonly resist direct oxidation by molecular oxygen which remains in a stable triplet ground state. The triplet state of oxygen ($^3O_2$) has two unpaired electrons with the same (paraller) spin in each of the two highest orbitals. The singlet state of oxygen ($^1O_2$) is one in which the valence electrons are in the same orbital and have opposing or anti parallel (↓↑) spin or no magnetic moment. Singlet oxygen is not a free radical but on electrophillic species requiring electrons to fill the vacant molecular orbital which is very unstable At the time and during senescence, univalent reduction of dioxygen gives rise to activated species, such as, superoxide (O^{-2}, HO_2), hydrogen peroxide (H_2O_2), hydroxy radicals (OH) and finally water according to the following scheme (Elstner, 1982):

$$O_2 \rightarrow O^{-2} \text{ or } HO_2 \rightarrow H_2O_2 \rightarrow OH + H_2O \rightarrow 2H_2O$$

Various substances may act as electron donors in the above reaction chain, viz., photosynthetic mitochondrial and microsomal electron transport system, substrates of auto-oxidizable enzymes (peroxidases) and organic molecules, reduced metal ions etc. Superoxide has more stability than singlet oxygen. Both of them become strong nucleoplic and a free radical by the addition of electron. The acidic form of superoxide is the perhydroxyl radical. Superoxide (O^{-2}) can oxidize molecules of biological importance like sulphur compounds, o-diphenols, ascorbate or NADPH as well as it can reduce cytochrome C, metal ions and metal complexes. Adding an additional electron to O^{-2} results in the formation of H_2O_2 or O^{-2}. Hydrogen peroxide in the considered to be stable under physiological conditions of pH and temperature and in the absence of scavengers such as catalase and peroxidase. It is a powerful oxidant in cellular oxidative metabolism the last step in the oxygen reduction series is the formation of hydroxyl radical (OH) which is a strong oxidizing agent and swiftly reacts with most organic compounds found in the cell. Its life time in a cell is very short.

There are various sources of activated oxygen which can react among themselves and the actual form of activated oxygen in any of the particular physiological system has been always in double. In the presence of light, O^{-2} or H_2O_2 is a by-product of photosynthesis (Mehler reaction). Hydrogen peroxide is the product of other oxidases including glycolate oxidase. Superoxide can dismutase spontaneously or with the aid of superoxide dismutase (SOD) from $^1O_2$ and H_2O_2, respectively.

Superoxide dismutases are present in all aerobic organisms and play an essential role in the survival of organisms in the presence of oxygen. They protect the tissues from the deleterious effects of the oxygen free radical O^{-2} (superoxide) formed in various enzyme reactions, as mentioned above, in which a single electron is transmitted to O_2. A superoxide dismutase (SOD) with one atom of manganese per enzyme molecule has been isolated in leaf extracts of pea. Manganese-containing SOD is not as widespread as the copper-zine-containing SOD. In contrast, the iron containing SOD is restricted only to a few plant families. In illuminated green cells, the chloroplasts are the organelles with the highest rate of oxygen turnover, including O_2 consumption and in leaves more than 90% of the SOD is localized in the chloroplasts and only 4-5% is in the mitochondria. The exact localization of the isoenzymes Cu-Zu-SOD and Mn-SOD within the chloroplast (whether in the storma or the thylakoid membranes) is yet not clear The severe disorgaization of the chloroplast lamellar system (photooxidation) that occurs when manganese is deficient strongly support the role of manganese in this protective system. It is not clear, however, whether this enzyme has a specific requirement for manganese o whether other divalent cations such as Zn- might be at least as effective under *in vivo* conditions. The copper-zinc-SOD has a molecular weight 32000 approximately, and at the active site probably one copper and one zinc atom are closely connected by a common histidin.

Free radicals participate chiefly in the form of activated O_2 species, such as O^{-2} H_2O_2, in several electron transfer reactions normal cellular metabolism. The activate oxygen may leak out from faulty defence mechanisms and, in addition, give rise to the generation of the, even more injurious and aggressive oxygen radicals, e.g., OH; O^{-2} lipid hydroperoxides which readily react with nucleic acids, proteins, lipids, chlorophyll etc accelerating senescence. Adverse effect on the physiological and biochemical activities of plants are often exerted by certain heavy metals like Pb^{2+} and Hg^{2+} as they increased permeability of tissue due to membrane damage in some aquatic plants partially reducing oxygen species are known cause membrane damage by deesterifying membrane lipids.

Cytosolic Ca^{2+} levels during senescence: Senescence in a deteriorative process and is associated with a wide range of biochemical changes. Most notable in green leaves is the decrease of Chl content, who is accompanied by a parallel degradation of polar ion from the matrix of the thylakoid membrane (Thomas as Stoddart, 1980; Thomas, 1986; Meir and Philosoph-Haes, 1995) and by proteolysis (Thimann, 1985; Philosoph-Haes *et al.,* 1994). One of the factors that may participate in regulation of leaf senescence is [Ca^{2+}]cyt (Leshem, 1987), the regulation of which is considered essential in all eukaryotes (Bush, 1995).

Cellular Ca^{2+} homeostasis is normally maintained at submicromolar levels by membrane associated Ca^{2+} transport pumps, secondary transporters, and ion channels (Hepler and Wayne, 1985; Bush, 1995). Stimulus-induced changes in [Ca^{2+}]cyt may be transient, sustained, or oscillatory, and the time required for full response varies from a few seconds to several hours (Bush, 1995) A transient increase in [Ca^{2+}]cyt may result either, from a decrease in plasma membrane Ca^{2+}, ATPase activity (Paliyath and Thompson, 1988) or form the opening of Ca^{2+} channels in the cell membranes (Schroeder and Hedrich, 1989).

Previous studies of various systems have indicated that [Ca^{2+}]cyt is involved in the regulation of senescence and ripening processes in plants. It was previously hypothesized (Leshem *et al.,* 1984; Leshem, 1987) that deterioration of the membranes during leaf senescence could be linked to phosphlipid catalytic processes induced by a Ca^{2+}-calmodulin complex. Further evidence for the involvement of [Ca2+]cyt in the processes leading to senescence was reported for leaves of cowpea (Savithramma and Swamy, 1989), oat (Dreier, 1990), rice (Huang *et al.,* 1990; Chow and Kao, 1992), corn (Huang and Kao, 1992a, 1992b), and parsley (*Petroselinum cripum*).

In addition, the loss of Ca^{2+} transport capability has been shown to be an early event in the senescent stages of carnation flowers, which is likely to result in higher than normal levels of [Ca^{2+}]cyt and, therefore, the facilitation of senescence (Paliyath and Thompson, 1988). In tomato and apple fruits stage-specific changes in Ca^{2+} -dependent protein phosphorylation, which disappear completely at the onset of ripening and senescence, have been observed (Poovaiah and Reddy, 1987).

Role of Ca^{2+} and Calmodulin Mediated Processes in Senescence

Studies on the role of Calcium in mediating diverse physiological and development processes of plants are now appearing frequently. Many new roles are assigned to this element.

The possible role of Calcium in plant senescence may be inferred from the fact that calcium is widely known to play a major role in membrane structure and function (Jones and Lunt, 1967). Its effects on the maintenance of RNA and protein have been well demonstrated (Treviwas, 1970; 1972) and these components are considered as the indices of senescence. Each of the parameters of senescence, such as loss of chlorophyll, decrease in protein and increase in apparent free space and hydraulic conductivity is affected by the calcium status in the tissue (Poovaian and Leopold, 1973; Poovaiah, 1987).

It has been shown that Ca^{2+} contributes to the coupling of stimulus to response in plants. This is evidenced by the fact that photoactivated phytochrome initiates Ca^{2+} fluxes in plant cells (Roux, 1984; Das and Sapory, 1985) and thus serves as a key element in the transduction pathway of phytochrome response. Recent investigations demonstrating changes in the intracellular calcium concentrations due to stimuli such as light and hormones have provided the basis for suggesting calcium as a second messenger (Roux and Slocum, 1982; Marme and Dieter, 1983; Veluthambi and Poovaiah, 1984; Poovaiah, 1985; Hepler and Wyne, 1985; Poovaiah and Reddy, 1987).

Calcium distribution at the cellular level plays a critical role in cell function. High level of free calcium in the cytoplasm is injurious to the cell as such all organisms maintain low concentrations of free calcium in the cytosol, usually less than 10^{-7}M (Clarkson and Hanson, 1980). Cytoplasmic free calcium in plants, as in animals is ordinarily in the range of 0.1 to 1 μM (Bush and Jones, 1986). Even when the cell is activated by influx of calcium, any structure in the cell that is to serve as a target of calcium dependent regulation must have a binding affinity for calcium of around 10^{-5} M. Moreover, since the concentration of free magnesium in the cytosol is relatively constant at 10^{-3}M, the calcium binding sites must have a selectivity over Mg^{2+} of atleast 1000 fold. Several Ca^{2+} binding proteins fulfill this condition.

The major carrier to the flow of information into the cell is the plasmamembrane, where transduction mechanisms translate external signals. A transient flux of calcium appears to be the trigger allowing chemical messages to be relayed to the biochemical machinery within the cell. Cells have evolved an elaborate system of proteins that interact with calcium, controlling the transmission and reception of the intercellular message. These interactions may lead to better control over the intercellular Ca^{2+} concentration, a possibility that has broad implications for the control of plant growth and development.

Calcium effects on various parameters of foliar senescence, such as loss of chlorophyll and protein, increased membrane leakage, dissolution of middle lamella, associated changes well established. As such the present study is aimed at understanding the Ca^{2+} and calmodulin related processes associated with senescence of primary leaves of cowpea.

The primary leaves of cowpea (*Vigna unguiculata* L.) start senescing from 15th day after sowing of seeds. The following parameters were studied on 7th, 12th, 22nd and 32nd days of leaf growth and senescence:

1. Changes in total chlorophyll and protein content phospho and glycolipids and the activity of phosopholipase, quantification of Ca^{2+} content of the whole leaf as well as in the organelles like chloroplasts.
2. Mitochondria and microsomes.
3. Ca^{2+} ATPase activity lipid peroxidation.
4. Superoxide dismutase and catalase activity.
5. Polyamines and argimine decarboxylase activity.

The results obtained in the present study are briefly summarised below:

- Calmoduline concentration for unit leaf area is estimated in all the reference stages along with levels of Ca^{2+} Calmodulin is identified, quantified and characterised, Mol. wt. is determined through electrophrosis and identification is carried out through phosphodiesterase activity.
- To characterise the specific nature of Calmodulin, amino acid analysis is carried out. An attempt is made to correlate the activities of SOD, catalase; Lipoxygenase, Phospholipids tp Ca^{2+} and calmodulin related events during natural foliar senescence.

- It is observed that Ca^{2+} and CaM levels per unit leaf area of the primary leaves showed unaltered concentration throughout the period of senescence.
- Lipid peroxidation, Lipoxygenase activity, Phospholipase activity showed an increase with the progress of senescence.
- Total protein content increased from 7th to 12th day and gradually decreased but calmodulin level remained constant.
- The ratio of calmodulin to total protein content per unit leaf area showed a decrease upto 12th day and followed by an increase till 32nd day. Ca^{2+} level also showed a similar trend.
- The unaltered levels of Ca^{2+} and calmodulin indicate their active role in the progress of senescence.

Mineral Uptake

Redistribution of minerals probably occurs to some extent in all multicellular plants and during many stages of development, but it is particularly prominent during fruit growth (Loehwing, 1951; de Mooy *et al.,* 1973; Ohlrogge, 1960; Williams, 1955). Mineral distribution in soybean has been studied most extensively (Borst *et al.,* 1931; Dermann *et al.,* 1978; Hammond, 1951; Hanway *et al.,* 1971; de Moody *et al.,* 1973). N, P, K, and Mg, but not Fe, Ca, S, and Mn seem to be redistributed from the leaves to the developing fruit of soybean. Mg, however, is not readily translocated.

Senescence is characterized by the transition of the metabolism from nitrogen assimilation to nitrogen remobilization (Feller and Keist, 1986; Peoples and Dalling, 1988). Nitrogen is redistributed in maturing cereals from senescing leaves to the maturing grains, allowing an efficient use of this frequently limiting macronutrient (Feller, 1990). The importance of nitrogen remobilization from senescing leaves for the nutrient economy is obvious.

The relative ability of developing seeds to mobilize different minerals has been compared in a single study employing three species of legumes including pea (Hocking *et al.,* 1977). The leaves, carpels, and seed coats lose 60 to 90 per cent of their N, P, and K to the seeds, but they release only 20 to 60 per cent of their Zn, Mn, Fe, and Cu less than 20 per cent of Ca and Na. There is no doubt that mineral may be taken out from senescing organs, particularly leaves, for use by developing fruit. On the surface, deficiency created in the donor organs provides a sufficient cause of their senescence.

According to Sonnewald and Willmitzer (1992) there are two types of sink organs: utilization sinks (e.g. growing organs) and storage sinks (e.g. tubers, seeds, storage roots etc). In addition to the above there may be another sink, known as alternative sink (Biswas and Mandal, 1987b) which is flexible and adjusts itself according to the nutritional status and changing developmental and source-sink relationship of the plant. Such sink organs may be stem, fruit wall, leaf etc, which may also serve as intermediary storage for the transfer of assimilates to the growing organs. There is no rigid rule as regards to the particular identity of source and sink organ. A leaf may change from net importer to net exporter during ageing and transport system is sufficiently flexible to that (Evans, 1975).

In soybean, reproductive sink removal (depodding) prevents mineral redistribution just as it prevents senescence (Derman *et al.,* 1978). It may be supposed that this would be true for monocarpic plants in general. Prevention of flower development in tobacco by cutting off the plant's top, however, does not completely prevent the loss of N from the lower leaves, for the N and is redistributed to support additional growth of the roots, stem, and uppermost leaf group (Watson and Petrie, 1940). Thus, in species where flowers or fruit removal reinstates vegetative growth, these new sinks can induce redistribution. Minerals are released from senescing leaves for use by developing fruit. Yet, discrepancies exist; for example, in soybean, foliar applications of complete mineral nutrient solutions do not retard senescence, even though enough minerals penetrate to increase seed yield (Derman, Rupp and Noodén, 1978). In addition, soybean can be completely defoliated as late as full bloom (at which time re-growth of the foliage still occurs), but seed yield is reduced by only 20% (Weber, 1955). This does not support the idea that fruit growth is totally dependent on a reservoir of nutrients that is accumulated in the leaves prior to the start of reproductive growth. Perhaps more important, manipulation of reproductive sink size (pod dose) in soybeans shows that the foliar senescence response is saturated at a relatively low sink size (<50% of the maximum), rather than showing a good parallel (Lindoo and Noodén, 1977; Noodén, and Derman Rupp, 1978).

In wheat there is a dependence of nitrogen transport on carbon transport to the ear (Barneix *et al.*, 1992) where about 80-90% of nitrogen is accumulated before anthesis (Cregan and Vanberkum, 1984), while most of the carbon is accumulated after anthesis. The altered source-sink relationship during reproductive development determines the onset of senescence of source organs in different ways.

Enzymes

There is loss of Rubisco activity in dark induced senescence of mustard cotyledons. The activity of protease increases during senescence, which increase degradation of protein. There is an increase in peroxidase during senescence, which takes part in degradation of lipid and bleaching or carotenoids. Decrease in the level of mRNA responsible for the synthesis of the small subunit of Rubisco during dark-induced senescence of mustard cotyledons. A temporal correlation between enzyme protein breakdown and senescence-induced activity of proteases may suggest proteolytic degradation of the enzyme. A senescence-induced protease capable of hydrolyzing soluble proteins, including Rubisco, has been demonstrated in *Festuca* chloroplasts. The protease is characterized by its ability to hydrolyze different substrates, including Rubisco protein, at different pH during senescence (Peoples *et al.*, 1980).

As far as enzymes involved in nitrate reduction and amino acid metabolism, the enzymes nitrate reductase and nitrite reductase are only needed for the assimilation of inorganic nitrogen. Nitrate reductase is located in the cytosol and requires NADH or NADPH for its function, while nitrite reductase is a chloroplast enzyme and depends on reduced ferredoxin (Wallsgrove *et al.*, 1979; Beevers and Hageman, 1980; Nelson *et al.*, 1984; Streit *et al.*, 1985).

Stomatal Conductance

In soybean, De Souza (1997) found that water stress condition declines in stomatal conductance. De-fruiting caused stomatal resistance (Koller and Throne, 1978; Setter *et al.*, 1980) to increase yet depodding in soybean inhibited rather than promoted senescence, along with the inhibition of transpirational flux and partitioning of nutrients.

Seed filling period and Yield

In soybean, De Souza (1997) found that water stress during seed filling period causes shortening the seed filling period. Water stress during seed filling could reduce soybean (*Glycine max* L.) yield by accelerating leaf senescence and shortening the seed filling period (De Souza, 1997). Water stress during seed filling reduces yield by reducing seed size (Ashley and Ethridge, 1978: Korte *et al.,* 1983a; Meckel *et al.,* 1984: Smiciklas *et al.,* 1989: Vieira *et al.,* 1992). These reductions in seed size have been associated with earlier maturity (Ashley and Ethridge, 1978) and a shorter seed filling period (Mackel *et al.,* 1984; Smiciklas *et al.,* 1989). It is generally assumed, that the shorter seed filling period results from accelerated leaf senescence (Sionit and Kramer. 1977: Egli and Carfts-Brandner, 1996), but direct measures of leaf senescence supporting this assumption are lacking.

Leaf senescence in soybean characterized by declines in leaf N, chlorophyll, and photosynthesis, begins early in seed filling and is usually complete by physiological maturity (Wittenbach *et al.,* 1980; Egli and Crafts Brandner, 1996). During leaf senescence, C and N are redistributed from the leaf and other vegetative plant parts to the developing seed. This redistributed N can be a significant source of N for the seed, contributing from 20 to 100% of the mature seed N in cultivars of differing maturity (Zeiher *et al.,* 1982). The amount of redistributed N in the seed and the N available for redistribution at the beginning of seed fill are directly related to the vegetative mass at beginning seed fill, which is largest in late-maturing cultivars (Zeither *et al.,* 1982). Similar results have been obtained when the variation in vegetative mass and redistributable N at beginning seed fill was created by water stress (Egli *et al.,* 1983).

Water stress during seed filling could reduce yield by accelerating the decline in leaf photosynthetic activity and increasing the remobilization of C and N to the seed.

Chapter 5

PHYSIOLOGICAL AND MOLECULAR ASPECTS REGULATING SENESCENCE

Senescence is a complex, highly regulated, developmental phase in the life of a leaf that results in the coordinated degradation of macromolecules and the subsequent mobilization of components to other parts of the plant. Senescence of leaves is not simply concerned with death, but that leaf senescence represents a key developmental phase in the life of both annual and perennial plants, which is as ordered and complex as any other phase of development. Indeed, leaf senescence is as important a developmental phase as those of growth and maturation. During its life span, a leaf undergoes at least three phases of development. Initially, it is expanding rapidly, importing carbon and nitrogen and undergoing rapid protein synthesis until its full capacity for photosynthesis is reached. Then the mature leaf becomes an asset to the plant, contributing to the supply of carbon, during which time, protein turnover is at a consistently low level. This continues until internal or external conditions initiate the onset of senescence. Leaf senescence is a period of massive mobilization of nitrogen, carbon and minerals from the mature leaf to other parts of the plant and is a highly regulated, ordered series of events involving cessation of photosynthesis, disintegration of chloroplasts, breakdown of leaf proteins, loss of chlorophyll and removal of amino acids. From studies with plants and animals, a number of theories have been proposed to explain aging processes, mainly those that are degenerative in nature. They are as follows:

- There is an accumulation of toxic or deleterious products within the organism.
- There is a loss of some essential substance from the organism.

- The organism undergoes general *wear and tear,* a process that damages or destroys the integrity of the sub-cellular organization or metabolic apparatus.
- Through mutations, certain cells accumulate deleterious genes (genetic damage).

However, these theories are not mutually exclusive; and different organisms may show different aspects of ageing and senescence. Apart from above, following theories have been proposed by different physiologists. These are as follows:

NUTRITIONAL THEORY OF SENESCENCE

The German physiologist H. Molisch suggested that senescence might be caused by nutritional deficiencies. The general similarities between monocarpic senescence and reproductive processes led Molisch (1938) to postulate that senescence of aged organs is produced by a competence between these and the reproductive structures for nutrient uptake. At the end of this competence between the vegetative and reproductive structure for nutrients uptake, the nutrient drawn from the vegetation to reproductive. Organ, resulting in exhaustion of nutrients, hence they die. It is not always true that distribution and redistribution of nutrients are the regular features during senescence. Therefore, during senescence, the leaves of many species export to other structures of the plant N, K and in some cases Mg. Whereas senescent leaves import from other structures of the plant massive amount of Pb, Zn, Cd, Cu, Ni. Mn, Al. Cr, Na, and Ca. Probably depending upon the ions present in the plant, senescence is also a mean to eliminate ions in excess. This function would be important for the elimination of toxic elements out of the plant, which moreover would make difficult the growth and competence by plants in the soil near to the plant with senescent leaves.

Nutrient removal by flowers or fruits is not the entire explanation of senescence, even for soybeans, because the young flowers could not possibly divert enough nutrients to cause death of leaves. In cocklebur plants, short-day and long-night conditions induce flowering and leaf senescence, but even if all flower buds are removed, leaf senescence still occurs. Furthermore, development of male flowers on staminate spinach plants induces senescence as effectively as development of both flowers and fruits on female plants, even though staminate flowers divert far fewer nutrients that do fruits and seeds.

Kelly and Davies (1988) concluded from an extensive literature review that diversion of assimilates to developing fruits is no longer accepted by most researchers as the strongest regulator of senescence. They proposed instead that development of the reproductive phase itself somehow causes diversion of nutrients to flowers and fruits, slowing of vegetative growth, and then the later stages of senescence. They emphasized that as reproductive structures become strong sinks, vegetative organs somehow also become strong sinks, vegetative organs somehow also become weaker ones. In roots, loss of sink strength is accompanied by decreasing transport of mineral nutrients and cytokinins upward through the xylem. It is quite likely that a decreased cytokinin supply to leaves is partly responsible for the start of leaf senescence.

HORMONAL THEORY OF SENESCENCE

Noodén, Lindoo and Leopold have shown that Molisch's theory was insufficient to explain senescence fully. According to them although nutrient redistribution may be a function of senescence and it is not necessarily the cause of senescence. For instance, deflowering or depodding usually retards senescence, however, this is not always the case especially with reference to defloration in Locklebur or de-fruiting in corn and pepper. These conditions do not retard (in some case they do accelerate) senescence.

It must also be noted that nutritional competence can hardly explain the senescence preceding the fall and death of the leaves of deciduous trees. If it is possible that plant have evolved mechanism for the control of senescence by hormones, possibly they have mechanism for the hormonal control of their correlative development e.g., it is probably that in a deciduous plant such as spinach the male plant has no nutritional mobilization for monocarpic senescence. Thomas and Stoddort (1980) were of the view that probably hormones along with nutritional factors control senescence, thereby showing that hormones do play as an active role in senescence.

Death Hormone Hypothesis and Senescence

The term death hormone has been used by Noodén and Leopold (1978) when discussing senescence in plants that flower only once (monocarpic senescence), e.g., pea, soybeans, cereal grains, some bamboo species etc. In most cases, but not all, senescence is inhibited or retarded if the plant is prevented from flowering or setting fruit. This can be done by keeping the plants at non-inductive day lengths and temperatures, or by manual deflowering and de-fruiting (Noodén and Leopold, 1978). Two theories have been proposed in relation to monocarpic senescence. These are:

Death hormone hypothesis

This explains that senescence factors are produced by the flower, fruit, seeds which are transported to the vegetative parts (Noodén and Leopold, 1978). They induce cessation of growth, degradation of cell apparatus and remobilization of nutrients to the maturing seeds.

The nutrient drain hypothesis

This hypothesis proposed by Molisch explains that there is a pull from the sink of young reproductive tissue monopolizes all nutrients, helped by a reduced sink strength of the vegetative parts in the flowering.

Characteristics of Death Hormone

Death hormones are produced in developing seeds (Noodén and Leopold, 1988)

They seem to be transported back to the plant in the xylem (Noodén and Murray, 1982; Noodén and Leopold, 1988)

They have limited mobility in the xylem. A little more in the phlóem (Noodén and Leopold, 1988)

In soybean they act primarily on the leaves, in pea on the apical meristems (Gianfangna and Davies, 1983; Noodén and Leopold, 1988)

It is usually agreed that plant hormone groups influence senescence. In most plants cytokinin are anti senescence compounds that can postpone plant death; gibberellins also have similar effects, but in fewer plant species. Auxin influence senescence and abscission in very complex manners, apparently depending upon the auxin source, or depending on the age and receptivity of the tissue. When applied to the leaf, auxin often retards senescence, but paradoxically, when auxin enters the leaf from the stem, it sometimes promotes senescence.

Ethylene and abscisic acid are the two important candidates for death hormones. Ethylene is closely involved in various senescence phenomena, especially flower wilting and fruit ripening. Though no direct evidences are available regarding the involvement of ethylene in monocarpic senescence, however, one could imagine that the ethylene precursor, aminocyclopropane carboxylic acid is transported, and ethylene is a kind of secondary messenger (may be Ca-calmodulin) within the individual plant tissue. There is little correlation between abscisic acid (ABA) concentration and plant death.

GENETIC REGULATION AND MOLECULAR BIOLOGY OF SENESCENCE

Many workers like Woolhouse (1976), Thomas and Stoddort (1980), Leopold (1980) considered senescence as a genetically programmed process, which culminate the development and differentiation of plant system. Senescence is an ordered series of cytological and biochemical events. On the cytological level, some organelles are destroyed while others remain active. The chloroplast is the first organelle to deteriorate during the onset of leaf senescence, with the destruction of thylakoid protein components and stromal enzymes (Hensel et al. 1993). In contrast, nuclei remain structurally and functionally intact until the late stages of senescence. Senescing tissues carry out catabolic processes that require the *de novo* synthesis of various hydrolytic enzymes, such as proteases, nucleases, lipases, and chlorophyll-degrading enzymes. The synthesis of these senescence-specific enzymes involves the activation of specific genes.

PCD refers to a process by which cell promote their own death through the activation of self-destruction systems. It is now generally accepted that many plant developmental processes and stress responses are achieved through the operation of PCD; these include senescence morphogenesis, embryogenesis, sex determination in monoecious plants such as maize, abscission zone formation, and hypersensitive reaction to pathogen infection (Green, 1996; Jones and Dangl, 1996). The concept of leaf senescence as a type of PCD has been recognized for many years (Noodén and Leopold, 1978). However, it is unclear whether leaf senescence shares any biochemical or genetic pathways with other types of PCD in plants or animals.

Not surprisingly, the levels of most leaf mRNAs decline significantly during the senescence phase, while the abundance of certain other transcripts increases. Genes whose expression decreases during senescence are called **senescence down-regulated genes (SDGs)**. SDGs include genes that encode proteins involved in photosynthesis (Hensel *et al.,* 1993; John *et al.,* 1995), in accordance with previous studies that demonstrated deterioration in photosynthetic activity and the sequential disappearance of chloroplast components (Bate et al. 1991; Hensel et al. 1993). However, senescence involves much more than the simple switching off of photosynthesis genes. Many genes whose expression is induced during leaf and flower senescence have been identified through differential screening of cDNA (complementary DNA) libraries (Grbic and Bleecker 1995; Drake et al. 1996).

Genes whose expression is induced during senescence are called **senescence-associated genes (SAGs)**. The functions of most SAGs have been deduced from sequence homology to known genes, and confirmed by their enzyme activities. Among them are genes that encode hydrolytic enzymes, such as proteases, ribonucleases, and lipases (Hensel et al. 1993; Taylor et al. 1993; Ryu and Wang 1995; Oh et al. 1996), as well as enzymes involved in the biosynthesis of ethylene, such as ACC (l-aminocyclopropane-l-carboxylic acid) synthetase and ACC oxidase (O'Neill et al. 1993). SAGs of another class have secondary functions in senescence. These genes encode enzymes involved in the conversion or remobilization of breakdown products, such as glutamine synthetase, which catalyzes the conversion of ammonium to glutamine and is responsible for nitrogen recycling from senescing tissues (Watanabe et al. 1994).

The leaf senescence program is accompanied by changes in gene expression. This was first demonstrated using in vitro translation followed by gel electrophoresis to detect changes that occur in translatable mRNA populations during leaf senescence (Watanabe and Imaseki, 1982). Analysis of the in vitro translated proteins revealed that an abundance of most leaf mRNAs significantly diminshed during the progression of senescence, whereas some translatable mRNAs increased during senescence (Watanabe and Imaseki, 1982; Davies and Grierson, 1989; Becker and Apel, 1993; Buchanan-Wollaston, 1994; Smart *et al.*, 1995).

Differential screening of cDNA libraries made from mRNAs of senescent leaf tissues also demonstrated that the expression of the vast majority of genes is down-regulated, whereas the expression of other genes is up-regulated during senescence. For example, the abundance of transcripts encoding proteins involved in photosynthesis decreases sharply during senescence (Bate *et al.*, 1991; Hensel *et al.*, 1993; Jiang *et al.*, 1993; Humbeck *et al.*, 1996); however, mRNA levels of certain genes increase with the progression of leaf senescence.

To date, about 30 SAGs (senescence-associated genes; defined as genes with expression up-regulated during leaf senescence) have been identified from a variety of plant species, such as *Arabidopsis* (Hensel *et al.*, 1993; Taylor *et al.*, 1993; Lohman *et al.*, 1994; Oh *et al.*, 1996), *Asparagus* (King *et al.*, 1995), barley (Becker and Apel, 1993), *Brassica napus* (Buchanan-Wollaston, 1994), maize (Smart *et al.*, 1995), radish (Azumi and Watanabe, 1991), and tomato (Davies and Grierson, 1989, Drake *et al.*, 1996)

What triggers a suicidal impulse?

Before giving the answer of this question, a dynamic cell system should be realized in terms of chemical kinetics. It is well known that the primary structure of a polypeptide chain is determined by the nucleotide sequence in a structural gene. A region called the operator is associated with structural genes and it initiates the transcription and synthesis of the corresponding mRNA. A number of structural genes may be associated with an operator. This complex is named as operon. A regulator gene is a distinct entity which has an association with an operon. A substance called repressor is produced from a regulator gene which has reversible combination capacity with corresponding operator. This results in the operator-repressor complex blocking the transcription off the structural gene. There is obvious reason to believe that repressors may be supplied from environment and the action of this repressor can be antagonized by an inducer. The above theoretically model can be realized biochemically in a cell and can change its course of action at will, but a lot remains to be gained from such a genetically controlled mechanism. However, this model gained evidence from many developmental problem of senescence has become prohibitively difficult.

A model (Strehler, 1969) endeavoured to present a fairly practical theory. The idea started by taking into consideration two main mechanisms in a cell i.e. control of message synthesis and control of message translation. It is well known that all amino acids are coded in several different ways by different codones in the DNA and that the translation capacity of a code in different cell types is basically varying. Therefore, the synthesis of a protein in a given cell is determined by various messages and is limited by the translation capacity of a cell. It is to be noted here that the production of any specified protein will be prevented if a single code is left untranslated. Thus, it appears that each developmental stage is activated by synthesis of a new species of amino-acylated tRNA molecules and the repression of this activation in a cellular pool may perhaps, be due to a loss of specific amino-acylated tRNA. This led Strehler to believe that senescence sets in as consequence of an original loss of specific amino-acylated tRNA species. He put forward evidence for the fact that different cell types of the same system lack different tRNA acylation enzymes, which leads to suggest that some control at least, can be exercised at the translational level in cell differentiation and senescence.

Two molecular approaches to the causal mechanism of senescence are worth mentioning. Norman and G. Anderson (1960-62) stressed upon the defects arising in the genetic code and accumulation of defective non-mobile biosynthates due to environmental inflictions. Strehler (1969) considers that living body is nothing but a correlated complex of numerous enzymes. Absence of availability of a single amino acid or non-coding of even a single codon can render this enzyme ineffective. As specific aminoacylated transfer RNA molecules maintain the availability of amino acid to respective codon, presence or absence of a specific amino acylated soluble RNA will consequently induce or repress the synthesis of enzyme.

Woolhouse (1987) postulated that at the onset of senescence a receptor system is activated at the cell surface or in the cytosol and the signal is conveyed to the nucleus by intermediary messengers. As a result, the pattern of nuclear gene expression changes and signal reaches the chloroplast via nuclear-encoded proteins which cross the plastid envelope. Whatever signalling processes are involved, the primary inducer of senescence may ultimately affect the amount of or activity of transcriptional factor that interact with cis-elements of senescence associated genes (SAGs). Reversible protein phosphorylation is the major mechanism by which the conformation and activity of proteins including transcriptional factors is modulated in eukaryotic cells.

The primary inducer activates the synthesis and release of short-lived second messengers, intracellular signalling molecules that bind to and activate ion channels and enzymes including protein kinases (PKs) and protein phosphatases (PPs). These enzymes may activate transcription directly or indirectly through their action on other enzymes via translocation to the nucleus following activation or via activation of cytoplasm located transcriptional factors which subsequently relocate to the nucleus.

The signals that trigger senescence in plants are unknown. However O_2^- and H_2O_2 accumulate in senescing tissues (Pastori and del Rio, 1997). It is possible that reactive oxygen species are the general trigger for senescence in plans. An NADPH oxidoreductase is thought to generate $O_2^-$and H_2O_2 and it may be activated after translocation of the putative regulatory subunits to the plasmamembrane (Xing *et al.*, 1997). It is not known how certain cells in plans die, whereas others often very close to the dying cells continue to live. One possibility is that the generation of O_2^-, H_2O_2 or other molecules that trigger programmed cell death may be controlled at a supracellular level and accumulate only in cells that are fated to die. Another possibility is that the expression patterns of catalase and peroxidase genes which can also be regulated at supracellular level (Willekens *et al.,* 1994), are controlled to degrade H_2O_2 and prevent programmed cell death in cells that live.

Calcium ions also take part in signalling system. Ca^{2+} are released through the action of the primary inducer of senescence and bind to calmodulin which activates phospholipase A_2. This enzyme releases linoleic and linolenic acids from membrane phospholipids. These fatty acids are precursor of Jasmonic acid. Jasmonic acid and other products of this pathway, which include ethylene and Ca^{2+} inophores may recycle to the membranes, allowing more Ca^{2+} ions to enter and generating a cascade effect.

Whatever initiates senescence must have the capability to activate senescence specific gene expression. However, it is not known at what level of gene expression this activation may take place. None of the senescence – associated genes identified up to now appears to be a regulatory gene that has function in initiation or control of senescence (Buchanan-Wollaston, 1997).

Genes Expressed During Leaf Senescence

In the last few years, enabled the isolation and characterization of a large range of cDNA clones representing genes that show increased expression in senescing leaves. The analysis of these genes and identification of the function of the encoded proteins will allow a picture of the complex processes that take place during senescence to be assembled. Once of cDNA clone that detects an increased transcript level in senescing tissue has been identified, Northern hybridization can be used characterize the expression of the gene in terms of steady-state mRNA levels. To data, genes encoding degradative enzymes such as proteases and nucleases, enzymes involved in lipid and carbohydrate metabolism and enzymes involved in nitrogen mobilization have all been identified; their role in senescence may be less predicatable and, possibly, more interesting.

Leaf senescence does not occur by a passive decay mechanism, but rather, is an actively regulated process that involves coordinated expression of specific genes. During senescence, the levels of total RNA decrease and the expression of many genes is switched off (Bate *et al.,* 1991: Henesl *et al.,* 1993; Lohman *et al.,* 1994). It has been postulated for some time that the senescence process may depend on *de novo* transcription of nuclear genes and recent molecular studies have shown that this is the case. New transcripts were originally detected as a changed pattern of products observed in *in vitro* translation experiments using mRNA from senescing leaves when compared to green leaves (Malik, 1987; Thomas *et al.,* 1992). Recently, using differential screening and substrative hybridization techniques, many cDNA clones representing senescence-enhanced genes have been identified from a range of different plants and these are summarized in Table 5.1. The identification of genes that are expressed specifically during senescence is the best evidence that de novo transcription is required for senescence to proceed.

Table 5.1: Cloned genes that show increased levels of transcription during senescence

Gene name	Possible function	Class	Plant
SAG2	Cysteine protease		*Arabidopsis*
Seel	Cysteine protease	5	Maize
LSC7	Cysteine protease	7	*Brasscia napus*
See2	Cysteine protease	5	Maize
SAG12	Cysteine protease	5	*Arabidopsis*
LSC790	Cysteine protease	10	*Brasscia napus*
LSC760	Aspartic protease	7	*Brasscia napus*
UB17	Polyubiquitin		Potato
RNS2	Ribonuclease		*Arabidopsis*
MS	Malate synthase	6	Cucumber
ICL	Isocitrate lyase	6	Cucumber
PBPCK-7A	Phosphenolpyuvate carboxykinase	6	Cucumber
gMDH	NAD-malate dehydrogenase	6	Cucumber
LSC101	Fructose 1.6 bisphosphate aldolase	10	*Brasscia napus*
LSC540	Glyceraldehyde-3 phosphate dehydrogenase	10	*Brasscia napus*
See3	Pyruvate *o*-phosphate dikinase	10	*Maize*
LSC213	Pyruvate *o*-phosphate dikinase	5	*Brasscia* napus
PTIP11	β-galactosidase	10	*Asparagus*
PLD	Phospholipase D		Castor bean
Atgst2	Glutamine synthetase	7	*Arabidopsis*
	Glutamine synthetase		Radish
	Glutamine synthetase		Rice
LSC460	Glutamine synthetase	7	*Brasscia napus*
PTIP12	Asparagine synthetase	9	*Asparagus*
LSC54	Metallothionein I	5	*Brasscia napus*
JET12-like	Metallothionein		Sambucus
LSC210	Metallothionein II	7	*Brasscia napus*
RgMT	Metallothionein II		Rice
LSC680	ATP sulphurylase	9	*Brasscia napus*
GSTII-27	Glutathione S-transferase	5	Maize
LSC650	Catalase	7	*Brasscia napus*
LSC550	Cytochrome P450	9	*Brasscia napus*
LSC260	Cytochrome P450	9	*Brasscia napus*
LSC226	Cytochrome P450	5	*Brasscia napus*

Gene name	Possible function	Class	Plant
LSC94	Prla	8	*Brasscia napus*
LSC222	Chitinase	5	*Brasscia napus*
LSC212	Antifungal protein	7	*Brasscia napus*
TOM13	AGC oxidase	5	Tomato
TOM75	MIP membrane channel		Tomato
LSC8	NADH: ubiquinone oxidoreductase	10	Brasscia napus

Protein Degradation: Involvement of Genes

The degradation of proteins is probably the most significant break down process that takes place during senescence since the remobilization of amino acids is very important to supply the developing organs elsewhere in the plant. Genes encoding several different types of protease have been identified in senescing leaves from various plants (Table 5.1). One of the most obvious enzymatic events that occurs in the senescence process is proteolysis and it is therefore not surprising that there is *de novo* transcription of protease genes and synthesis of protein. However, increased levels of specific proteases in senescing leaves have been hard to detect reproducibly by biochemical methods due to the high vacuole of the developing leaf (Huffaker, 1990; Feller and Fischer, 1994). The identification of senescence-enhanced protease genes is, therefore, significant. Much of the protein within a photosynthesizing cell is located within the chloroplast and it is likely that the initial steps in protein degradation occur within these organelles. Chloroplasts do contain proteases; protein degradation can be detected in isolated intact chloroplasts (Mitsuhashi and Feller, 1992) and ClpP and ClpC protease subunits have been detected in the chloroplasts of *Arabidopsis* (Shanklin *et al.,* 1995). However, senescence-specific protease activity that is involved with the dismantling of the photosynthetic apparatus within the chloroplast, remains to be demonstrated.

There are senescence-enhanced protease genes, isolated from maize, *Arabidopsis* and *B. napus* show sequence similarity to a class of cysteine protease represented by seed-specific proteases from cereals such as the oryzain Y protease from rice (Watanabe *et al.,* 1991). These enzymes are produced in germinating seeds and, perform a supply developing to seedlings. They are likely to perform a similar function in senescing leaves. Three other types of cysteine protease showing enhanced levels in senescing leaves have been identified.

One of these, represented by the cDNA clone SAG12 appears similar in protein sequence to papain-like proteases (Lohman *et al.,* 1994). This gene is expressed specifically in senescing leaves and, as is described below, work with the promoter of the gene has shown in to be tightly regulated, during leaf development (Gan and Amasino, 1995). In contrast, the third cysteine protease that has been identified in *Brasscia napus*, represented by cDNA clone LSC790, is expressed at all stages of leaf development and also in flowers and germinating seeds (Buchanan-Wollaston and Ainsworth, 1997). The level of transcripts of this gene is relatively high in young green leaves, decreases in mature leaves and increases significantly during senescence. The most similar sequence to the protein encoded by the LSC790 gene is a drought-induced cysteine protease, RD19, that has been isolated from *Arabidopsis* (Koizumi *et al.,* 1993). The *Brasscia napus* gene represented by LSC 760 which encodes an aspartic protease showed a different pattern of expression; the gene was expressed similarly in young and mature green leaves but expression increased during senescence (Buchanan-Wollaston and Ainsworth, 1997).

One of the genes shown to be senescence enhanced in maize encodes a protein similar to a vacuolar processing enzyme from castor bean seeds, which is involved with the conversion of vacuolar proteins to their mature forms (Smart *et al.,* 1995). The expression of this gene during senescence may indicate that some of the senescence-related degradative enzymes are synthesized in the green leaf, stored inactive in the vacuole, and activated by such processing enzymes once senescence is in progress.

The location of the proteases that are synthesized during senescence remains to be determined. The protein sequence deduced from the two cysteine proteases and the aspartic clones identified in *Brassica napus* indicates that they all have similar hydrophobic N terminal regions which are likely to direct the protein into endoplasmic reticulum. None have obvious chloroplast transit peptide sequences which might indicate their localization in the chloroplast. However, the protein sequence derived from the cloned gene can be used to identify antigenic peptide sequences; this has been done for the LSC7 protein ad the peptides synthesized have been used to raise. Preliminary immunolocalization experiments with senescing *Brassica napus* leaf tissue has indicated that the protease encoded by LSC7 may be located within the chloroplast in senescing leaves (Fife and

Buchanan-Wollaston, unpublished results). Further experiments such as this with other senescence-related proteases will help to elucidate the problem of how and where protein degradation takes place during leaf senescence.

The ubiquitin pathway for targeted protein degradation is important in the elimination of abnormal cytosolic proteins and also in the rapid turnover of short-lived proteins. Ubiquitin is a highly conserved protein that is found in all eukaryotes (Callis and Vierstra, 1989).

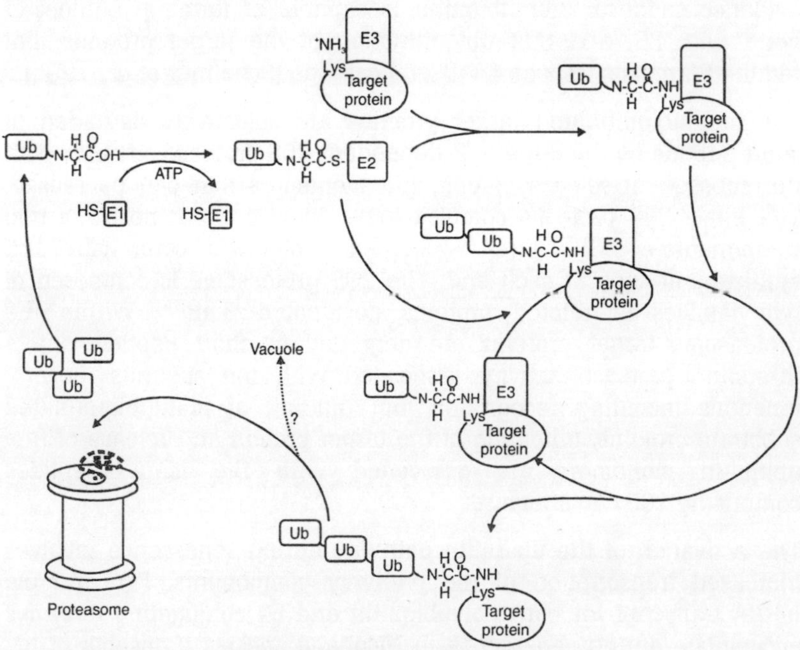

Fig. 5.1: The ubiquitin pathway (after Blanknap and Garbarino, 1996).

The recent biochemical and molecular evidence indicate that proteins are targeted for degradation by complexing with ubiquitin during senescence. The ubiquitin pathway facilitates bulk protein degradation for nitrogen recycling. The ubiquitin-dependent proteolysis of proteins occurs in two steps-ubiquitination of the NH_2 group of an internal lysine residue of the target protein followed by degradation in the proteasome (Fig 5.1). The covalent attachment of a ubiquitin monomer to the target protein requires three distinct enzymes, ubiquitin-activating enzymes (E1), ubiquitin conjugating enzyme (E2) and ubiquitin – protein ligase (E3).

The ubiquitination is initiated by ATP-dependent activation of ubiquitin catalyzed by E1. The ubiquitin is then transferred to an E2 carrier protein which can either function alone in target protein ubiquitination or act as a donor to the E3 – target protein complex (Ciechanover, 1994). Although E3 enzymes have been largely viewed as docking proteins required for substrate protein recognition, recent evidence suggests that ubiquitin is capable of forming a thioester bond with E3, and that ubiquitination of the target proteins that require E3 results from an E1-E2-E3 cascade (Scheffner *et al.,* 1995).

The multiubiquitinated target proteins are selectively degraded to short peptide by the large ATP-dependent 26S protease complex with the subsequent release of ubiquitin monomers that can be reused. The intact 26S protease complex found in vivo is composed of two components-a 20S barrel – shaped proteolytic core with 16S multisubunit caps at each end. The 20S proteasome is composed of two families of related proteins designated Band . Within the proteasome target proteins are degraded to short peptides via a threonine protease activity associated with the subunits. Several functions including recognition and binding of multiubiquitinated substrate proteins, unfolding of the target protein and release of free ubiquitin monomers are associated with the many peptides comprising 16S cap structure.

The activation of the ubiquitin pathway during senescence involves increased transcription of the pathway components. For example mRNA transcript for both polyubiquitin and E2 conjugating enzymes accumulate during senescence in Nicotiana species (Genschik *et al.,* 1994). Transcriptional activation is also suggested by experiments using transgenic potato plants in which a β-glucuronidase (GUS) maker gene is transcribed from a polyubiquitin promoter. Expression of the marker gene was five to ten-fold higher in senescent leaves relative to young leaves. The response of polyubiquitin gene expression to treatment with phytohormones is also consistent with upregulation during senescence.

Genes Involved in Nucleic Acid Breakdown

The nucleic acids, in particular the rRNA, within a senescence cell are an important source of carbon, nitrogen and especially phosphorus.

The level of DNA in a senescing cell remains relatively constant as senescence progresses, but the levels of ribosomal RNA decrease (Makrides and Goldthwaite, 1981). RNase activity in leaves has been shown to increase during senescence (Green, 1994). Three RNase genes that show increased expression under phosphate-limited growth conditions have been identified in *Arabidopsis.* One of these genes, RNS2, is also expressed at high levels during senescence and is likely to be important for the degradation of RNA and remobilization of Pi in senescing leaves (Tayor *et al.,* 1993; Bariola *et al.,* 1994). The fate of the purines and pyrimidines resulting from nucleic acid degradation is not clear, but they are presumably broken down further, to conserve their nitrogen and carbon components for export. The increased levels of uricase and xanthine oxidase observed in the peroxisomes of senescent leaves may indicate that purine catabolism takes place in this organelle (Vicentini and Matile, 1993; Pastori and del Rio, 1994).

Genes Involved in Lipid Remobilization

The membranes of plant cells constitute a valuable store of lipid molecules which can be mobilized and used by the senescing leaf. The level of total lipids decreases in senescing leaves and it appears the membranes of the cell including the thylakoid membranes are metabolized to provide energy for the senescence process (Koiwai *et al.,* 1991; Wanner *et al.,* 1991). Respiratory activity in a senescing leaf is high a continuous energy supply is necessary to allow degradation and mobilization processes to take place. Meanwhile, the supply of photosynthate to support respiration is decreasing due to dismantling of the chloroplasts. There are several lines of evidence that suggest that membrane of ^{14}C-labelled oleic acid, incorporated in galactolipids by feeding to greening etiolated barley, is released mostly as CO_2 during senescence (Wanner *et al.,* 1991). The respiratory quotient (RO) of senescing leaves tends to be low, implying that the substrate for respiration is fatty acids rather than sucrose. Also, levels of key enzymes of the glyoxylate cycle, malate synthase and isocitrate lyase have been shown to increase in senescing leaves of barley (Gut and Matile, 1988). The glyoxylate cycle, originally identified in micro-organisms as the pathway allowing growth on acetate, was shown to operate in higher plants in endosperm tissue of castor oil seeds (Kornberg and Beevers, 1957). The cycle is involved with the conversion of acetyl CoA, produced after lipid breakdown, to four carbon acids for subsequent conversion to carbohydrates, particularly sucrose, by the reversal of glycolysis, or gluconeogenesis.

The presence of the pathway is characterized by two key enzymes, malate synthesized in cotyledons during post germinative growth when stored lipid is being mobilized to provide energy for growth (Comai *et al.,* 1989). The importance of the glyoxylate cycle in senescence has been confirmed by the detection of increased expression of the genes encoding these two enzymes during leaf senescence in cucumber (Graham *et al.,* 1992; McLaughlin and Smith, 1994).

Therefore, it appears that cellular lipids, such as the galactolipids, which are the main component of the thylakoid membranes, or the endomembrane lipids, may be metabolized during senescence via the glyoxylate cycle (Vicentini and Matile, 1993; McLaughlin and Smith, 1995). Sucrose can be synthesized from the four carbon products of the glyoxylate cycle via the process of gluconeogenesis, which converts TCA cycle components to sucrose via a reversal of the glycolytic from the senescing leaf. The co-ordinate expression of the gene encoding phosphoenolpyruvate carboxykinase, a key enzyme involved in the conversion of lipids to sugars, with the genes for malate synthatase and isocitrate lyase during the natural senescence of cucumber cotyledons, adds weight to the suggestion that the expression of these genes is related to gluconeogenesis (Kim and Smith, 1994a). PEP carboxykinase catalyses the conversion of oxaloacetate to PEP, an essential step for gluconeogenesis to occur Similarly, increased expression of peroxisomal NAD malate dehydrogenase, one of the enzymes involved in the glyoxylate cycle, has been detected in senescing cucumber cotyledons (Kim and Smith, 1994b).

The senescence-related increase in expression of genes that encode enzymes involved in the glycolytic pathway may also be indicative of the importance of gluconeogenesis as a senescence-related process. Genes encoding enzymes such as fructose, 1,6, bisphosphate, aldolase and glyceraldehyde-3-phosphate dehydrogenase are expressed in young *Brassica napus* leaves, expression is lower in mature green leaves and then increases again during senescence. These enzymes may be involved in supporting the increased respiration observed during senescence or, may be important for the synthesis of sucrose from components of degraded lipids and proteins, via gluconeogenesis. Interestingly, another gene that is expressed during senescence is that encoding pyruvate orthophosphate dikinase. Senescence-induced expression of this has been observed in maize (Smart *et al.,* 1995) also in *B. napus* This enzyme is a key component of C_4 photosynthesis where its role is to synthesize phosphoeolpyruvate (PEP) from pyruvate to restore PEP levels for CO_2 fixation. The gene encoding this enzyme is expressed at high levels in young developing leaves in maize (a C_4 plant): expression is decreased in mature leaves and then increases again during senescence. In *B. napus* (a C_3 plant), very low levels of expression are detectable in young and mature green leaves and expression increases during senescence. The implication of this is that pyruvate orthophosphate dikinase enzyme, in addition to its role in C_4 photosynthesis, has an alternative role to play during senescence in both C_3 and C_4 plants. The most likely role for this enzyme is in the gluconeogenesis pathway, where it may be involved in the conversion of pyruvate, produced from amino acid breakdown, into PEP.

Recently, senescence-related expression of a gene encoding phospholipase D, an enzyme that may be involved in the hydrolysis of membrane phospholipids, has been shown in dark-induced senescence of castor bean leaves (Ryu and Wang, 1995). This enzyme may catalyze the first degradative step leading to the deterioration of membrane integrity during senescence (Thompson *et al.,* 1987). In addition, the enhanced level of expression of a ß-galactosidase gene that has been detected in senescing asparagus ferns may be involved with the mobilization of galactose during degradation of the galactolipids which are the major components of the thylakoid membranes (King *et al.,* 1995).

Genes Involved in Chlorophyll Breakdown

Leaf yellowing, the visible sign of senescence, is due to preferential degradation of chlorophyll over caroteoids (Matile, 1992). The fate of Chl during senescence has successfully been elucidated in the course of the last decade. Intermediate and final catabolites have been structurally analyzed and the enzymes of the catabolic pathway are biochemically well characterized. Despite these efforts, some questions or different aspects of Chl breakdown remain unanswered the factor(s)initiating and mechanisms regulating Chl degradation are elusive up to nm, attempts to purify enzymes of the pathway have resulted in proteins senesce information only in few cases. Further biochemical work, but also molecular strategies will be necessary to eventually clone the respective genes. Projects on functional genomics and introgression mapping analysis of stay-green phenotypes will be helpful in that respect as well. It is not overly speculative to suggest that within the next decade there will be another major step forward in slowing the biological enigma of Chl breakdown (Hortersteiner, 1999).

Genes Involved in Nitrogen Remobilization

Degradation of macromolecules during senescence is followed by remobilization of the components to developing areas of the plant. The nitrogen that is present in protein and nucleic acid molecules is converted to transportable amino acids, in particular the amides, glutamine and asparagine, which are the predominant amino acids in the phloem transported from senescing leaves (Kamachi *et al.,* 1992; Feller and Fischer, 1994).

It is likely that, during senescence, ammonia is released by deamination of amino acids and catabolism of nucleic acids and is converted into glutamine by the enzyme glutamine synthetase (GS). There are two distinct forms of GS found in plant leaves-GS1 which is located in the cytosol and GS2 which is plastidial (Kamachi *et al.,* 1991). During senescence, the activity of GS2 decreases while that of GS1 tends to increase (Kamachi *et al.,* 1992). The senescence-related increase of expression of genes encoding GS1 has been observed in several different plants (Table 5.1) which demonstrates the importance of GS during senescence and confirms the prediction of its role in the conversion of the nitrogen from macromolecules into glutamine. The expression of GS1 appears to be mainly confined to the vascular bundles where it is, presumably, involved in the generation of glutamine for transport, while GS2 is found in photosynthetic tissues where its main role is in the re-assimilation of photorespiratory ammonia (Kamachi *et al.,* 1992). Therefore, the increase in expression of the GS1 gene that is observed in senescing leaves is likely to be due to an increased requirement for glutamine synthetase to transport the mobilizable nitrogen away from the senescing asparagus ferns leaf (King *et al.,* 1995).

The source of the ammonia and glutamine for the synthesis of glutamine synthetase is likely to be through transamiation and deamination reactions involving amino acids released from the degraded proteins (Feller and Fischer, 1994). The amide group from most amino acids can be transferred to α-ketoglutarate by the action of a transaminase to produce glutamate, and glutamate dehydrogenase may then be involved in the production of ammonia from glutamate (Smart, 1994). Other sources of ammonia may be from the breakdown of other macromolecules such as nucleic acids. The source of the α-ketoglutarate, a component of the TCA cycle may be maintained by substrates supplied from the glyoxlate cycle or the transamination reactions will liberate a variety of 2-oxo-acids that can enter the TCA cycle and be used either for further trasamination reactions. Amino acids such as threonine and serine cannot normally be metabolized by transamination; a senescence-specific threonine dehydratase that has been identified in tomato may be involved in the release of ammonia from these amino acids during senescence (Szamosi *et al.*, 1993).

A summary of the metabolic pathways that may be involved in the remobilization of macromolecules during senescence, leading to export of nitrogen as the amides, gluconeogenesis. The expression of genes encoding enzymes of the glyxoylate cycle is rare in plant cells, occurring only in germinating seeds where oil stores are remobilized for respiration, and in senescing leaves where it is likely that it functions to mobilize membrane lipids to use in respiration or to synthesize sucrose which is exported from the senescing cell. The succinate produced from the conversion of acetyl CoA in the glyxoylate cycle can feed into the TCA cycle where it can be converted to oxaloacetate for gluconeogenesis or a-ketoglutarate for transamination to glutamate. The nitrogen component of the mobilized amino acids is probably converted via glutamate to glutamine by glutamine synthetase and some may be converted to asparagine by asparagine synthetase for export. A proportion of the glutamate may be converted back to α-ketoglutarate by glutamate dehydrogenase to provide ammonia for the glutamine synthesis. The carbon skeletons from the amino acids can enter the gluconeogenesis pathway as pyruvate, acetyl CoA or a-ketoglutarate. The senescence-related is probably involved with the conversion of the pyruvate, produced from amino acid breakdown, into PEP to enter into the gluconeogenic pathway.

Senescence-related Genes of Unknown Functions

The senescence-enhanced genes discussed so far, encode enzymes that appear to be involved in the degradative and remobilization processes of senescence. In many cases, increased levels of the encoded enzymes have been detected during senescence and the enhanced expression of these genes could have been predicted. However, it is not as easy to postulate a senescence-related function for some of the other genes that have been identified.

Senescence-enhanced expression of metallothionein-like genes has been detected of a gene encoding ferritin has been found to increase during senescence. Both metallothioneins and ferritin function by binding metal ions-iron ions in the case of ferritin and a range of divalent metal ions in the case of metallothionein (Theil, 1987; Kagi, 1991). The possible functions of this type of protein in senescence are not clear. It is possible, that, during leaf senescence, the release of metal ions from the protein breakdown that occurs is sufficient to warrant a detoxification role. Alternatively, these metal ions may form a valuable resource to the future development of the plant and so the presence of a metal binding protein may have a function in the storage, ad possibly transport of metal ions. There is some evidence that metabollothioneins may act in mammalian systems to protect DNA from oxidative damage caused by free radicals (Chubastu and Meneghihi, 1993). Leaf senescence is an oxidative process; breakdown of chlorophyll and membranes causes an increase in free radical production. Therefore, the presence of metallothionein proteins may protect the nuclear DNA from damage, thus allowing expression of senescence-specific genes required for the process to take place.

During senescence, it is essential that the senescing cells remain viable for as long as possible to allow maximum exploitation of their cellular components. Therefore, it is likely that there will be mechanisms in place to protect the cells from damage by internal or external factors. Levels of free radicals and reduced oxygen species such as hydrogen peroxide increase in response to environmental stress and also during senescence (Merzlyak and Hendry, 1994). The plant responds to the presence of these toxic compounds by the synthesis of a range of antioxidative enzymes such as superoxide dismutase. ascorbate peroxidase and catalase (Foyer *et al.,* 1994). Catalase is located in peroxisomes which become converted to glyoxosomes in senescing leaves (Vicentini and Matile, 1993). The function of catalase, acting in parallel with ascorbate peroxidase located in the chloroplast, is to detoxify hydrogen peroxide that is produced during photorespiration in photosythetic cells and also during degradation of macromolecules during senescence (Foyer *et al.,* 1994). Catalases are present at all stages of leaf development of and three catalase genes showing differential expression have been identified in *Nicotiana plumbaginifolia* (Willekens *et al.,* 1994). A catalase gene has been identified in *Brassica napus* that shows high levels of expression in late senescence (Buchanan and Ainsworth, 1977). The enzyme encoded by this gene may function to protect the senescing cell from damage at later stages of senescence, when, possibly, the chloroplast-located ascorbate peroxidase system becomes less active.

Other genes showing senescence-related expression may be involved in the metabolism of sulphur. A *Brassica napus* gene encoding ATP sulphurylase shows enhanced expression during senescence (Buchanan-Wollaston and Ainsworth, 1997). This enzyme activates sulphate in the presence of ATP in the step in the pathway to cysteine and methionine biosynthesis (Schmidt and Jager, 1992). Free cysteine is converted to the tripeptide glutathione which appears to have an important role in the response of a plant to many external and internal stresses (Wingate *et al.,* 1988; Foyer *et al.,* 1994). Glutathione may act as storage molecule for sulphur and this reaction might be important for the mobilization of sulphur released during macromolecule degradation from the senescing leaf. The gene encoding glutathione S-transferase also shows increased expression in senescing leaves (Smart *et al.,* 1995). This enzyme is involved with detoxification of herbicides by conjugation with glutathione and the senescence-related expression may be related to stress. Alternatively, it could have some role in sulphur metabolism during senescence.

Three different cytochrome P_{450} genes, that are expressed in senescent and not green leaves, have been identified in *B. napus.* In plants, cytochrome P_{450}s are key enzymes involved in the synthesis of a large variety of secondary plant metabolites including flavenoids, phytoalexins and lignin (Butt and Lamb, 1981). The role of these enzymes in leaf senescence is not understood. Another class of gene that has been found to show increased expression in *Brassica napus* senescence is that encoding proteins normally associated with the pathogen response (Hanfrey *et al.,* 1966). These genes include PRIa, a chitinase and a gene showing similarity to an antifiungal protein. It is possible that the presence of the PR proteins in senescing tissues reflects the vulnerability of the tissues and performs an ancillary function as part of the defense system against invading pathogens. Senescing leaves, containing a relatively high concentration of breakdown products and having a weakened physical structure, are extremely susceptible to pathogen attack. Alternatively, these genes may play fundamental roles in senescence since their role in protection against pathogens is not clear. It is possible that the expression of some of the PR proteins could be concerned with the induction and control of leaf senescence and only indirectly with defense against pathogens.

To date, only few SAGs are senescence specific. These include SAG12 and SAG13 from Arabidopsis (Gan, 1995) and L5C54 from Brassica napus (Buchanan-Wollaslon, 1994). Both SAG12 and SAG13 are senescence specific and are expressed in other senescing green tissues and floral organs such as stems, sepals, petals and carpels. The patterns of SAG expression change in response to different treatments or conditions. For example, cDNA clones isolated from dark induced senescing barley leaves is not expressed during natural senescence. SAG12 which is expressed during natural senescence is not detectable in leaves during the initial stages of dark, ethylene or ABA induced senescence. Also, the expression of individual SAG genes may also change in response to different stimuli. These results indicate that there may be multiple senescence regulatory pathways that activate distinct sets of genes and some genes may share common pathways. (Gan and Amasino, 1997). SAG promoter sequence analysis has revealed that the regulation of SAG expression is multifactorial and there is no common cis-acting element in their promoter region (Oh *et al.,* 1996).

Genes Expressed in other Degradative Physiological Processes

There are similarities between many of the events that take place during leaf senescence and fruit ripening, for example, chlorophyll degradation, increased activity of hydrolytic enzymes and, in some cases, ethylene synthesis. The involvement of genes common to the two processes might, therefore, be expected. Several cDNA clones representing genes that show increased expression during leaf senescence were identified by screening clones originally shown to be expressed during tomato fruit ripening (Davies and Grierson, 1989). One of these genes, TOM13, encodes the ethylene biosynthetic enzyme ACC oxidase (or EFE) and antisense expression of this gene was found to delay leaf senescence in transgenic tomato plants (Picton *et al.,* 1993). Another of the ripening-related genes shown to be expressed in senescing leaves was TOM75 which encodes a membrane channel protein that may be involved in the transport of small molecules from senescing cells (Fray *et al.,* 1994).

Other physiological processes that involve metabolic events similar to those that occur during leaf senescence include petal senescence and post-harvest deterioration. Genes that have been identified showing induced expression in these systems show common functions to those identified in senescence. For example genes involved in ethylene biosynthesis are expressed in senescing carnation flowers (Park *et al.,* 1992) and in broccoli after harvest (Pogson *et al.,* 1995a). Protease genes are expressed in senescing ovaries in pea (Granell *et al.,* 1992) and in the petals of daylily and carnation (Jones *et al.,* 1995; Valpuesta *et al.,* 1995). A gene encoding a delta 9 desaturase, possibly involved in the degradation of membrane fatty acids has been shown to be expressed in senescing rose petals (Fukuchi-Mizutani *et al.,* 1995). In many cases, genes that have been identified as being involved in one of these physiological processes have also been found to be expressed in an alternative processes (e.g. fruit repining and leaf senescence [Davies and Grierson, 1989]; post-harvest degradation and leaf senescence (King *et al.,* 1995).

Identification of genes that show induced expression during leaf senescence and analysis of possible functions of the proteins encoded by these genes have not, so far, contributed very much to our current knowledge about the processes that occur during senescence. The reasons for this are that, in general, the cloned genes that can easily be assigned functions have already been implicated in senescence from enzyme studies. Transgenic plant analysis using the genes that, from comparison with homologous genes, have no obvious function in senescence or which no similar sequences in the databases, will be more likely to generate new information on the senescence processes that has not been accessible by biochemical studies. Also, the identification of more senescence-related genes will help to fill in the picture of the senescence process that is starting to emerge. The genes idontified so far, however, do open new ways for studying the regulation of senescence.

REGULATION OF LEAF SENESCENCE

The events that take place during senescence appear to be highly coordinated to allow the maximum use of the cellular components before necrosis and death occurs. Complex interactions involving the perception of specific signals and the induction of cascades of gene expression must occur to regulate this process. Expression of the genes coding for senescence-related proteins in likely be regulated via common activator proteins that are, in turn, activated directly or indirectly by hormonal signals. Little is known about the mechanisms involved either in the initial signalling or subsequent coordination of the progenes. However, the identification of senescence-related genes provides the tools required for the molecular analysis of the factors involved in the regulation of senescence.

There are two main approaches that may be taken for the analysis of transriptional regulation of gene expression. One of these involves the identification and analysis of mutants that are defective in the process of interest, an approach which can be powerful if a large number of mutants can be identified reasonably easily. For example, mutants defective in ethylene responses can be identified at the seedling level and dissection of the ethylene response pathway using mutants in *Arabidopsis* is well underway (Zarembinski and Theologis, 1994; Ecker, 1995). Senescence is a complex process involving diverse signals and the identification of regulatory mutants would require the characterization of a large number of potential senescence mutants. So far several groups have expended considerable effort in trying to identify *Arabidopsis* mutants defective in leaf senescence, with very little success (Bernhard and Matile, 1994).

The alternative approach to analyse the regulation of leaf senescence is to study the transcriptional activation of senescence-related genes. Cloning and characterization of the promoters of genes that show induced expression during senescence will enable the critical elements that are involved in regulating senescence to be defined. Even if the genes under study are not the primary sites at which senescence regulation occurs, it should be possible to trace back the regulatory steps to the first step in the pathway. In this way, the initial signals that regulate the onset of senescence can be defined. Of the many senescence-related genes that have been identified, Northern analysis has indicated that the transcription of the genes is not induced at the same stage of senescence. Therefore, the regulation of gene expression during senescence may involve several different signalling pathways; which are themselves coordinately regulated, activating transcription at different senescence stages.

Recent progress has been made in the molecular biology of leaf senescence. Many senescence – associated genes have been identified from several plant species. The lack of common sequence elements in the promoters of several senescence associated genes suggests that the regulation of senescence associated gene expression is multifactorial. It is postulated that there may be several genetic pathways that are interconnected to form a senescence regulatory network. The elimination of the activity of one gene in the regulatory network may have little effect on the overall progression of leaf senescence. The promoter elements that confer senescence specific expression and the transeriptional factors that interact with these elements need to be defined. Mutant screen is necessary which will identify genes that play a role in the regulation of senescence. Once the regulatory mechanisms underlying leaf senescence are understood, it should be possible to devise more sophisticated ways to manipulate leaf senescence for agricultural application.

Genetic Regulation of Leaf Senescence by Cytokinin

Treatment with cytokinins has been shown to delay leaf senescence in many plants (Van Staden *et al.,* 1988). Cytokinins are synthesized mainly in the root and transported to the rest of the plant in the xylem. It has been shown that the level of cytokinins in xylem sap declines when senescence is initiated and it has been suggested that a reduced level of cytokinins in the leaf may cause the onset of senescence (Noodén *et al.,* 1990).

Transgenic plants have been produced that have altered levels of cytokinins due to expression of an isopentyl-transferase gene (IPT, the enzyme that catalyses the rate-limiting step in cytokinin biosynthesis). Early experiments where this gene was expressed from a constitutive or heat shock promoter resulted in plants which did show delayed onset of leaf senescence, but the specific effects of the cytokinins on leaf senescence were hard to interpret because the plants grew abnormally (Smart *et al.,* 1991). Recently, the IPT gene has been cloned under the regulation of the promoter for SAG12-a senescence-specific *Arabidopsis* gene, encoding a protease (Lohman *et al.,* 1994) Transformed tobacco plants carrying this gene fusion developed normally up to the senescence stage, whereupon senescence was significantly retarded in the leaves from the transformed plants (Gan and Amasino, 1995). The leaves of the transgenic plants showed no signs of senescence, the photosynthetic activity remained high and more flowers were produced. The most important component of this experiment was the availability of the senescence-specific promoter for the SAG12 gene which is expressed only during leaf senescence and, therefore, has an auto-regulatory role. Expression of the promoter is induced as senescence starts, cytokinins are synthesized, senescence is halted, and the SAG12 promoter switched off. Thus, there is no cytokinin overproduction to affect the normal development of the plant (Gan and Amasino, 1995).

The control by cytokinins is likely to be at the transcriptional level- the presence of cytokinins above a certain level inhibits the expression of senescence-related genes. A fusion between the SAG12 promoter and the GUS reporter gene, when transformed into tobacco showed extremely high levels of the GUS protein in senescing leaves (Gan and Amasino, 1995). However, when this fusion was present in the same plant as the SAG12 promoter IPT gene fusion, the levels of GUS seen were much lower, indicating that the cytokinins produced by the IPT fusion were inhibiting the expression of GUS from the senescence-enhanced promoter. Two cytokinin-repressed genes have been identified in cucumber cotyledons (Teramoto *et al.,* 1995). The expression of these genes in mature green leaves is repressed in the presence of externally applied cytokinins. The level of transcript for both genes in very low in young leaves and increases significantly in mature and senescing tissue, presumably due to reduced levels of cytokinins present in this tissue. Therefore, it is clear that cytokinins, either directly or indirectly via a signalling pathway, can inhibit the transcription of senescence-related genes.

Genetic Regulation of Leaf Senescence by Ethylene

Ethylene plays a regulatory role in many plant processes such as seed germination, seedling development, fruit ripening, and flower senescence. Ethylene is also involved in plant responses to external signals, for example, responses to wounding, pathogens and stress caused by environmental pollutants such as ozone (Zarembinski and Theologis, 1994). In climacteric fruits, ethylene promotes ripening by co-ordinately inducing the expression of a large number of genes encoding enzymes required for the ripening process (Theologis, 1993) and in the senescence of some flowers, ethylene appears to have a similar regulatory role (Borochov and Woodson, 1989). Ethylene has been implicated in the regulation of leaf senescence in certain plants but, as is described below, recent evidence with transgenic plants and ethylene response mutants has indicated that, although ethylene has an effect on senescence, it is not an essential regulator of the process. In many plants, ethylene appears to have no role in ripening or senescence; for example, in the ripening of fruits of non-climacteric plants such as strawberry, in the senescence of some plants, in particular the monocots, there is no requirement for ethylene signalling (Smart, 1994; Valpuesta *et al.,* 1995).

Ethylene biosynthesis from S-adenosyl methionine involves two enzymes, 1-amino-cyclopropane-1-carboxylic acid (ACC) synthase and ACC oxidase. Increased expression of the genes encoding these two enzymes has been detected in ripening fruit, senescing petals and during post-harvest degradation of broccoli (Hamilton *et al.,* 1990; Park *et al.,* 1992; Pogson *et al.,* 1995b). In senescing tomato leaves, increased expression of the ACC oxidase gene was detectable early in senescence, before chlorophyll loss had commenced (John *et al.,* 1995). Experiments with transgenic plants that were expressing the antisense of the ACC oxidase gene showed that when the expression of this gene was repressed ethylene production reduced there was a marked effect on the rate of fruit ripening and a smaller effect on leaf senescence in these transgenic plants showed that senescence was delayed 10-14 d in the transgenic plants and the leaves remained photosynthetically had started, the progression of senescence in the leaves of the transgenic plants matched that taking place in the wild-type leaves (John *et al.,* 1995).

Mutants of tomato that are defective in the ethylene signal transduction pathway have been used to study the effects of ethylene insensitivity on fruit ripening and senescence. For example, plants carrying the Never Ripe mutation, which has been shown to block ethylene perception, produce fruit that ripen extremely slowly. The effects of the Never Ripe mutation have been studied in some detail by Lanahan et al. (1994). The flowers produced by these mutants do not senesce and show considerable delay in abscission. However, the leaves on the mutant plants, apart from showing a small delay in the onset of senescence, senesced normally. In parallel experiments, transgenic tomato plants overproducing ACC synthase were analysed; these plants produced high levels of ethylene which resulted in premature senescence of the flowers, but increased rates of leaf senescence were not observed. In fact, leaves on the ethylene overproducing plants remained green longer than control leaves; this lack of senescence may be related to the lack of fruit set due to flower abortion.

Experiments with the ethylene-insensitive *Arabidopsis* mutant, etr1-1, showed that leaf senescence was similarly delayed: leaves on mutant plants had an approximately 30% longer life span than the wild-type leaves although the photosynthetic activity of these longer lived leaves declined (Grbic and Bleeker, 1995). Once initiated however, the rate of senescence and the levels of expression of senescence-induced genes (SAGs) was found to be similar to that of the wild type. Treatment of wild-type plants with exogenous ethylene resulted in premature senescence with earlier expression of the SAGs although the increased expression was age related. Only the older mature green leaves showed a response to ethylene by inducing expression of SAGs, the younger, but still mature green leaves showed only marginal increases in SAG expression. This effect is comparable to results with tomato fruit where immature tomato fruit fail to ripen in response to ethylene and in appears that the tomato fruits have to be of a certain age before the ripening process can be induced by ethylene (Grierson and Kader, 1986). Also, ethylene could not induce the expression of ripening specific mRNAs in immature tomato fruit (Lincoln and Baker, 1987). These results have led to conclusion that other factors are involved in the regulation of gene expression in both fruit ripening and leaf senescence (Grbic and Bleeker, 1995) and that these 'age-related factors' and ethylene while for leaf senescence to occur 'age-related factors' only are necessary. The role of ethylene in leaf senescence appears to be some kind of modulator in determining the timing of leaf senescence.

Ethylene treatment of young *Arabidopsis* leaves does not result in increased expression of senescence-related genes (Grbic and Bleeker, 1995) although the expression is induced in the same tissue (Chen and Bleeker, 1995). Thus, transcription of senescence-related genes is not activated by ethylene unless the leaf is 'prepared' to senesce, in which case senescence is enhanced and three is concomitant expression of senescence-related genes. Therefore, the ethylene signalling pathway is probably not involved directly in regulating the transcription of genes during leaf senescence. The mechanism(s) by which ethylene modulates the rate of senescence is unclear, but it may be that the presence of ethylene increases the sensitivity of an inter-action that takes place at some point in the signalling pathway that is controlled by the presence of 'age-related factors'. In the absence of ethylene, transmission for a higher concentration of 'age-related factors'.

As well as hormonal signals, alterations in leaves of metabolites may be involved in regulating expression of some of the genes involved in the glyoxylate pathway that show senescence-induced expression are also regulated by levels of carbon compounds (Graham *et al.,* 1994; Kim and Smith, 1994a). It is possible that changes in levels of metabolites, that occur due to the reduced rate of photosynthesis, may have stimulatory effects on the transcription of senescence-related genes (Hensel *et al.,* 1993).

Thus, there appear to be least four signalling pathways that can be involved in the regulation of genes involved in leaf senescence.

Cytokinin levels

Senescence can only be inhibited when the cytokinins in the leaf fall below a threshold level. Maintaining the cytokinin levels inhibits transcriptional regulation of senescence-enhanced genes and prevents the onset of senescence.

Singals from developing sinks

Leaf senescence is repressed, or even reversed, when developing sinks are removed (Crafts-Brandmer, 1991). Tomato plants that overproduce ethylene produce no flowers or fruit and leaf senescence does not occur, possibly because of lack of signal from the developing fruit (Lanahan *et al.,* 1994). This signal does not overcome the cytokinin threshold signal since flower and seed development did not induce leaf senescence in the transgenic tobacco plants expressing IPT in the leaves (Gan and Amasino, 1995). However, in normal plants cytokinin levels may be influenced by the signal from developing sinks.

Ethylene

Antisense and mutant studies showing delayed leaf senescence have indicated that ethylene alters the timing of senescence. Increased levels of ethylene enhance the rate of senescence progression, but only in leaves that are already programmed to start senescence. Ethylene does not appear to activate senescence-related genes directly, but probably modulates the activation of the genes by other signals. Ethylene may have a role in repressing the expression of genes involved in photosynthesis (Grbic and Bleeker, 1995).

Level of photosynthate and other metabolites

The earliest sign of senescence initiation is the decline in carbon fixation rates which, in *Arabidopsis* and some other plants, commences as soon as full leaf expansion is reached (Hensel *et al.*, 1993). The decline in photosynthesis, resulting in reduced fixed carbon availability, may be a signal for the induction of expression of senescence-related genes.

These signals could act in concert to regulate the expression of senescence-related genes. The studies described here involve *Arabidopsis* and tomato leaf senescence. However, it is possible that the regulation of senescence in these two plants is not identical. For example, *Arabidopsis* leaf senescence does not appear to involve a signal from developing fruit (Hensel *et al.*, 1993) while that of tomato and other fruiting plants obviously does. Therefore, although recent results have shed some light on the regulation of leaf senescence, the developmental and physiological differences between different plants add a new level of complexity.

The isolation and characterization of promoters that control gene expression during senescence and the analysis of the factors that bind to these promoters and affect their expression during different developmental and environmental conditions may help to unravel the complexities of the regulation of leaf senescence as a whole. The promoter for the LSC54 gene, encoding a metallothionein-like protein, from *B. napus* (Buchanan-Wollaston, 1994), has been cloned and fused to the GUS reporter gene Analysis of the expression of this promoter GUS fusion in transgenic *Arabidopsis* has indicated that the promoter is highly induced during natural senescence in *Arabidopsis*. The promoter also shows increased activity in response to wounding and during dark induced senescence.

Interestingly, the expression of this gene in *Arabidopsis* appears to be liked to the disease resistance response to the pathogen *Peronospora parasitica*. The disease resistance response to pathogens in plants can often take the form of the hypersensitive response (HR) in which an infected plant reacts to an invading pathogen by killing the cells in the specific area of infection. This reaction of the plant has been likened to the programmed cell death phenomenon that has been studied in animals (Bowen, 1993; Mittler and Lam,1995). It has been shown that, in cells that are responding to a pathogen invasion by induction of the hypersensitive response, the promoter for the LSC54 gene is expressed very specifically, prior to cell death.

HR-associated cell death has been shown to require active plant metabolism and be dependent on *de novo* transcription and translation, that may point to HR-induced expression of the LSC54 gene, could imply that the signals involved in the disease resistance response, that allow the plant to respond to the presence of pathogen and prevent its spread, have elements that are common to pathways that are controlling cellular senescence. Further analysis of this promoter will allow the isolation of transcription factors and analysis of these should lead to the dissection of the regulatory steps that are involved in the senescence-related expression of this gene.

The application of molecular biology techniques to study the events that take place during leaf senescence has allowed considerable progress to be made in the last few years. Many senescence-related genes have now been identified and, although the functions of some of the encoded proteins have not been unexpected, transgenic plant analysis to identify the functions of other has begun to open new perspectives on the processes that occur during senescence. The availability of genes, promoters and mutants has allowed a clearer picture to be obtained concerning the regulatory mechanisms that are involved in senescence.

Chapter 6

ENVIRONMENTAL REGULATION OF SENESCENCE

Senescence is the last phase of leaf development, which is a process of degreening and produces changes at al levels of cellular and sub-cellular organization of leaves. Like many other, genetically programmed developmental processes, leaf senescence, particularly its initiation, is subject to regulation by many environmental and autonomous (internal) factors. The environmental cues include stresses such as extremes of temperature, drought, ozone, nutrient deficiency, pathogen infection, wounding and shading, whereas the autonomous factors include age, reproductive development and phytohormone levels. Once senescence is triggered, it is subject to environmental modulations and the reports on the effects of various environmental factors on leaf senescence are extensive (Biswal *et al.*, 1983b).

Environmental Cues and Senescence

The onset of senescence can not be detected easily as it is an integral part of leaf development rather than a dramatic change. The initial perturbation may cause only or small change in few control steps but result may be a cascade of secondary effects. Among the environmental cues, limited water and nutrient availability (especially nitrogen) are major factors that adversely affect plant life in many ecosystem. Plants have evolved mechanisms by which leaf senescence can be induced by these stresses to reallocate nutrients to reproductive organs and to eliminate water consumption by older, less productive leaves. This regulation of leaf senescence has an obvious adaptive value allowing the plant to complete its life cycle even under stressful conditions.

PHOTOREGULATION OF SENESCENCE

Light is the one of the important environmental factors, plays a very crucial role in photosynthetic green plants. In addition to its role in photosynthesis, it also decisively controls plant development and morphogenesis (Mohr and Shropshire, 1983). Light absorption by various photoreceptors, transmission of the signal and the final photoresponses in the greening process are well investigated (Kasemir, 1983). However, nothing definite is known about the mechanism of light action in controlling leaf senescence, although light has been known either to stimulate or retard the process (Maunders and Brown, 1983; Biswal *et al.*, 1983a, b, c; Panigrahi and Biswal, 1983; Satler and Thimann, 1983; Iordanov and M.G. Merakchilska-Nicolova, 1983; Rao, 1983; Schmutz *et al.*, 1983).

Photostimulation of Leaf Senescence

It has been known that photo-stimulation of chlorophyll loss during leaf senescence is mostly non-metabolic. The major photoreceptors involved are chlorophylls, which are photo-bleached at high intensity of light. This may cause a significant loss in photosynthetic efficiency of green leaves leading ultimately to their senescence. This type of senescence normally occurs in deciduous plants.

Maunders and Brown (1983) have shown light to degrade chlorophyll faster than dark, On the other hand, the differences between the leaves incubated in dark and light with respect to soluble protein, cytoplasmic RNA and free amino nitrogen are much less marked. These results would suggest that light action on chlorophyll degradation is non-metabolic senescence-associated changes. It is, therefore, important to differentiate between situations where light has got a direct photochemical role in the degradation of chlorophyll and light as a metabolic modulator of pigment degradation during leaf senescence. Whether light should act as a metabolic modulator or a photo-decomposer depends on the intensity of light and the plant system.

This has clearly been demonstrated by Pjon (1981) in leaf senescence of maize and hydrangea. He has shown the differential ability of leaf tissue of maize and hydrangea to dissipate their absorbed light quanta. The loss of chlorophyll in maize leaves has been attributed to the process of photo-oxidation of pigments, because continuous white light results in loss of maize leaf chlorophyll but retards the loss of protein. Photo-bleaching of chlorophyll in this system may be compared with the degradation of the pigments during aging of isolated chloroplasts (Raval *et al.,* 1982; Panigrahi and Biswal, 1983).

Reports of Lichtenthaler et al. (1982) provide new information on photostimulation of leaf senescence of seedlings grown in high and low light intensities. They have demonstrated that compared to low light strong, strong light (light quanta fluence rates) leads to the formation of sun type chloroplasts which exhibit high photo-synthetic rates and simultaneously an acceleration of leaf development, which also results in an earlier leaf senescence. Highlight intensities not only accelerate chlorophyll breakdown, but also lead to ultrastructural modification and loss in soluble carbohydrates. These changes, in addition to the breakdown of chlorophyll as shown by the authors, prevent a generalization that light stimulation of senescence is because of simple photooxidation of chlorophylls only.

Photoretardation of Leaf Senescence

The action of light in controlling leaf senescence is normally attributed to its effecting delaying senescence. It is to the economy of leaves that these should remain in a functional state of as long as possible once the energy has been invested in their formation. Therefore, it is logical to imagine the retarding effect of light in senescence once the leaves are matured. The reports on the precise role of light in retarding the process are controversial with mainly two different views, whether the action is photosynthetic or photomorphogenic (Biswal *et al.,* 1983b). This review is an attempt to report and retionalize the work from various laboratories engaged in this field. Since this is the first review in the field, some of the important references, although not very recent, are quoted.

Light Action Through Photosynthesis

The action of light in retarding senescence has been attributed by many authors to its role in photosynthetic production of organic nutrients. It has been demonstrated that application of exogenous sucrose to detached leaves incubated in dark delays senescence (Biswal *et al.,* 1983b). The participation of photosynthesis in regulating light-induced retardation of leaf senescence has been further confirmed by sensitivity of light response to DCMU, a photosynthetic inhibitor (Goldthwaite and Laetsch, 1967). More elaborative experiments on photosynthetic involvement in controlling senescence were conducted by Malik and Thimann (1980). Although they have shown that light action in retarding oat leaf senescence is not affected by DCMU, its response is significantly reduced when the leaves are treated with the uncouplers of photophosphorylation. They have, therefore, deduced that the involvement of photosynthesis in senescence regulation is primarily due to cyclic photophosphorylation, which, is not strongly affected by DCMU. On the other hand, Lloyd (1980) has demonstrated the importance of carbon dioxide in light-mediated retardation of senescence. He has shown senescence retardation in the leaves maintained in light with a normal atmosphere. Interestingly, light fails to retard senescence in the absence of carbon dioxide. This is supported by the recent work of Satler and Thimann (1983),who have shown that light without carbon dioxide does not have any senescence retarding effect and is comparable to the dark response. However, the nature of the complex interaction of light and carbon dioxide, whether photosynthetic or not, merits further investigation.

Action of Light Through Phytochrome

The first work on the non-photosynthetic action of light was reported by Haber et al. (1969), who have done extensive experiments on the senescence response to irradiance at various spectral zones of visible light and photosynthetic inhibitors. They concluded the role of light in retarding plastid senescence to be rather photomorphogenic. However, they did not demonstrate the precise nature of photoreceptors involved in the process.

Subsequently, the work conducted by De Greef et al. (1971) on the light retarding effect of senescence in *Marchantia* revealed phytochrome to be the photo-receptor, In recent days, many papers have appeared reported phytochrome as the major photoreceptor involved in light-mediated senescence retardation (Steinitz *et al.*, 1980; Pfeiffer and Kleudgen, 1980; Bucker, 1981; Biswal *et al.*, 1982; Biswal *et al.*, 1983). Steinitz et al. (1980) have investigated the effect of light in retarding senescence senescence of the flower stalk. Their results clearly suggest that the effect of light on retardation of chlorophyll degradation is mediated by the phytochrome system. Phytochrome has been shown to regulate plastid degradation during barley leaf senescence as reported by Biswal and Sharma (1976) and confirmed by Pfeiffer and Kleudgen (1980). The latter authors have shown that phytochrome not only retards chlorophyll loss but also the loss of carotenoids, plastoquinones and variable fluorescence of chloroplasts. Further, work on phytochrome as the photoreceptor in controlling leaf senescence of cucumber and tomato has been reported by Tucker (1981).

In these two systems, pulses of red light are shown to retard senescence approximately by the same extent as that of continuous white light and the effect of red pulses is nullified with brief irradiation of far red light after red light treatment suggesting phytochrome regulation of leaf senescence. In more elaborative experiments conducted by Biswal et al. (1982, 1983a) in mustard cotyledons, the effective control of phytochrome in retarding ultrastructure has been demonstrated.

Phytochrome as the real photoreceptor in light-mediated retardation of leaf senescence is further confirmed by the fact that a single RG-9 far red light pulse, which is ineffective with respect to photosynthesis, had a significant effect on the rate of senescence if the white light period is terminated by the far red pulse treatment before transfer of light grown seedlings to darkness. The treatment of the RG-9 far red light pulse at the end of a light period reverts nearly all the Pfr back to the Pr form (Biswal *et al.,* 1983a).

It is, at the moment, difficult to suggest any molecular mechanism of phytochrome action in delaying leaf senescence. However, control by phytochrome of cytoplasmic and plastid RNA during senescence of mustard cotyledons. The real story of phytochrome regulation of leaf senescence will be revealed with the experimentation on the action of the photoreceptor on leaf proteases, known to develop in dark-treated leaves during senescence.

Shding

Light flux decreases exponentially as it penetrates the canopy of leaves. It receives a lower ratio of red to far-red light as a result of preferential absorbance of red photons by chlorophyll in upper leaves. It has been shown that reduced PAR and a decreased red/far-red ratio are the senescence-triggering signals in shaded leaves. On the basis of these observations, Rousseaux et al. (1996). Suggested that a phytochrome signal pathway and photosynthate levels may be involved in triggering the leaf senescence programme. Transgenic plants that overexpress phytochrome A exhibit delayed senescence in shaded leaves.

Day Length and Light Flux

Day length and light flux appear to influence the onset of senescence in many species. Daylength may act through its effect on the switch to flowering which can trigger leaf senescence. Deciduous trees are sensitive to daylenth and temperature since a coordinated change in the colour of their foliage coincides with the shorter, cooler days of autumn. Titus (1989) reported that the first sign of leaf senescence in apple is a decline in proteins as the laylength becomes less than 14 h. When the daylength is decreased to less than 12 h, the leaves of chlorophyll, DNA and RNA begin to fall while RNase, polyphenol oxidase and malate dehydrogenase activities increase dramatically.

Indirect action of light

Action through photo-induced production of secondary messengers and other modifications of leaf physiology.

Light retarding action of leaf senescence is simulated in the laboratory by physiological modifications of the leaves, particularly by the treatment of leaves with exogenous ions, specific growth regulators in dark. It is, therefore, logical assume that light might be retarding senescence through its control over these factors in nature. In that case its action on senescence is rather indirect and through production of secondary messengers, a concept which has been proposed by many authors to operate during photomorphogenesis. These photo-effects, however, are reviewed subsequently either in terms of photosynthetic or photomorphogeic action of light.

Photoinduced changes in the level of phytohormones and senescence response. The literature on the response of leaf senescence to phytohormones is extensive (Woolhouse and Jenkins, 1983). Cytokinins, gibberellic acid and auxins are known to either inhibit or retard leaf senescence. On the other hand ethylene and abscissic acid are reported to promote the process. Light is known to be a decisive factor directing growth and morphogenesis including senescence of higher plants. But the precise relationship between light and phytohormones in leaf development is not know, although the involvement of light in hormones production is well established (Wareing and Thompson, 1976; Rohwer and Schierle, 1982). It is not unreasonable to assume that light could regulate the level of some specific plant hormones with a consequent effect on leaf senescence. In this regard, the work of Gepstein and Thimann (1981) has to be considered. They have shown that white light-induced retardation of oat leaf senescence is mediated through the light inhibition of ethylene accumulation. Wareing and Thompson (1976) have demonstrated a significant reduction in the level of endogenous ethylene with red light pulses, which on the other hand, stimulate the level of cytokinins. It is, therefore, possible that the effect of red light pulses could be simulated by the exogenous application of hormones like gibberellic acid and cytokinins. However, the interaction between phytochrome and exogenous hormones in controlling plant development seems to be complex in nature (Kasemir and Mohr, 1982). The effect of these two factors are shown either to be independent or additive. At present, there is no acceptable evidence that phytohormones could be perfect substitutes for light in photomorphogenesis.

Light-controlled polyamine level and senescence response. Polyamines, namely, putrescine, cadaverine, spermidine, spermineand their precursors like L-arginine and L-lysine are known to be potent inhibitors of leaf senescence (Galston, 1983; Muhitch et a., 1983). Interestingly, light which has been well known as a senescence retardant in excised leaves is also reported to increase the level of polyamines (Kaur-Sawhney *et al.,* 1982; Altman, 1982). Polyamine treatments of excised leaf segments of a number of plant species mainly retard in increase in activities of RNases and proteases consequently lowering the loss in leaf pigments, proteins and nucleic acids (Shih *et al.,* 1982; Altman, 1982; Kaur-Sawhney *et al.,* 1982; Galston, 1983). Polyamine induced senescence retardation through light may be interpreted in terms of inhibiting action of polyamines on the production of ethylene, a potent promoter of senescence (Evan-chen *et al.,* 1982; Apelbaum *et al.,* 1982). The next question arises as to how light regulates polyamine levels which consequently retard senescence? The precise mechanism of light action in polyamine metabolism is not clear. It is worthwhile to mention the work of Galston (1981) and Goren et al. (1982) who have attributed light action through phytochrome in controlling polyamine levels in pea plants.

Light mediated stomatal opening and senescence response: The opening status of stomata in recent years has been known to control light-mediated retardation of leaf senescence. The reports suggest light regulation of stomatal opening mainly by two different mechanisms. One is mediated by photosynthesis in the mesophyll cells. The control by light over the stomatal opening is, therefore, an indirect one. The other mechanism refers to the direct action of light on stomatal movement independent of photosynthesis. Senescence retarding action through the opening of stomata as controlled by light has been attributed by Thimann and Satler (1979) to the photosynthetic production of adenosine triphosphate (ATP). They suggest that the movement of the potassium ion in guard cells is dependent on ATP, a product of light reactions in photosynthesis. Subsequently, Gepstein and Thimann (1980) have demonstrated that the closure of stomata in dark results in an increase in the abscisic acid (ABA) level that accelerates the process of leaf senescence. Light fails to retard the increase in the ABA level it the stomata are artificially closed by floating the leaves on mannitol. However, recent findings of Satler and Thimann (1983) on the complex interaction of various atmospheric gases, namely, carbon dioxide, nitrogen and oxygen in the stomata exclude the possibility in generalization of a relation in photosynthesis, stomatal opening and senescence retardation.

In the second mechanism, the action of light in controlling stomatal aperture and its consequent effect on senescence retardation is attributed to the involvement of phytochrome and blue light photoreceptors (Morison and Jarvis, 1983; Nejidat *et al.*, 1983). It is known that stomatal opening is controlled by potassium ion flux which is dependent on ATP thought to be provided either by oxidative phosphorylations or photophosphorylations, which again suggests the participation of photosynthesis in stomatal opening. But it is possible that the amount of ATP required for stomatal opening could be provided by the action of phytochrome (Wagner, 1976). Secondly, an optimal pool of ATP is not the only factor for stomatal opening, since the exogenous supply of ATP in darkness does not show any effect on the status of stomatal movement as reported by Najidat et al. (1983). These authors have demonstrated that exogenous ATP controls stomatal movement in darkness only with brief irradiation of white or red light. They have suggested that of Mg^{+2} stimulated ATPase interacts with Pfr on membrane and thus controls stomatal opening.

However, the concept of light-mediated stomatal opening and its consequent retarding effect on leaf senescence has had a setback by the recent reports which do not suggest any direct correlation between the opening status of stomata and senescence regulation (Wardle and Short, 1983; Satler and Thimann, 1983; Biswal *et al.*, 1983c).

Light induced proton secretion and senescence retardation: Light has recently been shown to regulate senescence through secretion of protons from the leaf cells (Gepstein, 1982). It has been suggested that the proton pump in membranes eliminates protons froxm cytoplasm in the presence of light. This may prevent acidosis, consequently delaying the process of senescence by decreasing the activity of acid hydrolases.

It is logical to suppose phytochrome, a morphogenic photoreceptor, as the real photoreceptor to control leaf senescence rather than chlorophyll, the photoreceptor responsible for energy metabolism of green plants.

Secondly, if one assumes that light action in retarding senescence is due to production of secondary messengers, namely phytohormones and polyamines, then the existing data on photocontrol of these secondary regulators support phytochrome rather than chlorophyll as the photoreceptor. The current literature also suggests that light, through phytochrome action, brings about stomatal opening, a physiological modification reported to control leaf senescence, in a few plant systems.

The second aspect of light regulation of leaf senescence, which still demands attention, is the study on the precise effect of light in retarding leaf senescence: whether the maintenance of leaf is due to the formation of new organelles or retardation of disintegration of already formed ones. The existing reports suggests that light primarily checks the rapid breakdown of leaf organelles during senescence. This has been clearly demonstrated during plastid senescence of what (Haber *et al.*, 1969). The authors have shown that the effects of light in retarding senescence of chloroplasts is independent of its formation because light response with regard to senescence has been demonstrated in 3-amino-1,2,4-triazole (AT) treated leaves, in which plastic formation is prevented. This has been confirmed by Biswal et al. (1982, 1983a) during senescence of mustard cotyledons. In their experiments loss of chlorophyll has been chosen as a parameter to characterize leaf senescence. The advantage of using chlorophyll to monitor senescence is that its degradation after the transfer from light to dark could be studied without interference of its biosynthesis, since it is established that chlorophyll is not synthesized in most angiosperms without light. They have shown that continuous weak red light and red light pulses given every 12 h intervals are nearly as effective as strong continuous red light in retarding the loss of chlorophyll. Since the red light pulses do not contribute much to chlorophyll synthesis and their senescence retarding effects are fully reversible by far red light pulses, it is deduced that light acts at the level of chlorophyll degradation. Thus, with plastid degradation during senescence of mustard cotyledons as a model, we conclude that light controls leaf senescence primarily through phytochrome which, once the leaves are matured, retards their disintegration.

DISEASE-INDUCED SENESCENCE

The invading pathogen sets in motion a series of changes that reduce the functions (and "faculties") of the tissue. In short, the tissue senesces.

Pathogen Attack

Invasion by pathogens may retard or accelerate senescence. Rust and mildew infection of cereals typically cause the formation of green islands around infection sites. In contrast, localized mosaic or general yellowing are common symptoms of viral infection in leaves.

Effect of Pathogenicity on Plant in Relation to senescence

Photosynthesis

Since chloroplast show the earliest signs of damage, and their optimal activity requires that is coordinated with other cell compartments (cytokinin, peroxismomes, mitochondria), it is not surprising that in most discused tissues a decrease in photosynthetic activity has been found. Dissected leaves tend to be chloratic. However, not all cases of chlorophyll loss can be equated with disease-induced senescence. Perhaps the most con case of typical, disease-induced senescence is the yellowing of leaves infected by obligate parasites. Perhaps more accurately called parasites and/or viruses inducing local lesions. Small increases in photosynthesis have often been reported for the early phase of such infections but in the light of more recent studies it seems likely that the photosynthetic apparatus remains, at best, relatively intact and functional (Montalbini and Buchanan, 1974). The preservation of structure is logical since the biotrophic parasites require a continuous supply of metabolites for their own synthetic processes and even the from in which they receive these substances may be important.

Even the green island formation around the infection centers might represent a site of maintained photosynthetic activity rather than a site of increased chlorophyll synthesis and CO_2 fixation (Harding *et al.*, 1968). Also, high rates of incorporation into the macromolecular components in green islands do not necessarily indicate higher rates of synthesis but might reflect a change in endogenous precursor pools, altered turnover rates, and/or altered uptake processes. At any rate, a significant difference can be noted between the preservation of chloroplast structure and activity in green islands and the rapidly senescing parts of the same leaves. For example, one-half of a leaf moderately infected by rust accumulates more dry matter and grows more intensively than the slowly senescing non-infected half (El Hammady *et al.*, 1968). Maintained photosynthetic activity; cytokinin-directed, long-distance metabolite transport; inhibition of metabolite efflux; stimulation of RNA and protein synthesis; and inhibition of protein breakdown might equally play a role. The relative contribution of these factors will have to be elucidated.

Respiration

One of the most wider spread responses of host tissues to infection is an increase in respiratory rate. Such increase is not directly related to senescence. It can be detected relatively early infection. This is true for both virus-infected local lesion hosts and tissues attacked by biotrophic parasites. In this latter case the contributions of host and parasite are difficult to separate, the overall increase being very much confined to the infection centers.

The rise in respiratory rate reflects the general a anabolic nature of metabolism in infection centers. It is generally agreed that the preponderance of synthetic processes, both in host and parasite, and the associated high ATP-consumption, providing suitable phosphate acceptor (adenosine diphosphate) level, is the main controlling factor responsible for the high respiratory rates in tissues infected with biotrophic parasites. Especially during the sporulation phase there is a great demand for substrates and easily convertible energy. This requirement is met, at least partly, at the expense of non infected areas of the leaf or other parts of the plants, leading thereby to their premature senescence. These tissues are characterized by a low respiratory rate and loss of the respiratory control of their mitochondria.

Senescent tissues tend to be rich in peroxidases. Disease and damage add to the high peroxidase level. Although there is no proof for the in vivo functioning of peroxidases, at the later stages of infection and/or senescence the breakdown of cellular compartmentation might lead to peroxidative processes contributing to the overall O_2 uptake. Death is not an immediate event. Between a healthy and a totally collapsed cell a number of transition states occur with ample opportunity for the pathological operation of enzymes (peroxidase, polyphenol oxidase, ascorbic oxidase) with obscure in vivo function. These processes are, of course, secondary. Nevertheless, they can contribute to an autocatalytic speed up of aging and lesion formation, and can also affect symptom expression. The overall change can affect the spread of the parasite.

Respiratory increase in the infected host tissues is a common phenomenon with necrotrophic parasites as well and can be elicited by the application of pure toxins. Since toxin treatment results in a rapid loss of ions from the tissues, the idea of uncoupling of phosphorylation resulting in a lesion in energy-driven ion uptake and retention emerged. However, the uncoupling might be secondary since ion leakage precedes the onset of increase in respiration (Arntzen *et al.,* 1973). This aspect of parasitically induced respiratory increase has not been studied in relation to senescence.

Nucleic Acid Metabolism

To measure accurately the rates of synthesis of macromolecules like nucleic acids in higher plants is a difficult task because of the changing precursor pools and turnover rates. With fungal and bacterial diseases the extra precursor pools of the parasites complicate of the methodology. In *sensu stricto,* none of the works so far carried out on host-parasite complexes meets the present-day requirements. Nevertheless, at least with the biotrophic parasites, there are following evidences that at the infection centers nucleic synthesis is enhanced:

- Nuclei and nucleoli are enlarged, a phenomenon indicative of enhanced RNA (and protein) synthesis.
- Cytophotometric studies suggest an increase in RNA content within the nucleus and/or nucleolus.
- Biochemical analysis shows an increased RNA content in the infected leaf area.

Incorporation of labeled precursors into fungal as well as host cells is increased in, and in the vicinity of, the infection center (Bhattacharya *et al.,* 1965; Williams *et al.,* 1973).

Whatever the reason and mechanism of this increased synthetic capacity is, its consequences for leaf juvenility and/or senescence are obvious. If viewed under the electron microscope, the infected cell, in the early phase of disease, appears more juvenile than a mature one (Ehrlich and Ehrlich, 1971). The increased synthetic activity is, however, restricted to the infected area and its immediate vicinity. Later, even in the infected cell, the synthetic capacities start to decline but the parasite, especially in the compatible combinations, continues to accumulate RNA and proteins of its own. In this phase, the compartmentation of the metabolism of host and parasite becomes increasingly evident. Even the various compartments of the host cell behave differently. The chloroplasts show the first indications of damage. Correspondingly, the chloroplast RNAs start to incorporate less label from precursors, and their amount also decreases preferentially (Tani *et al.,* 1973a). This phenomenon is typical for aging, healthy leaves as well. Therefore, a careful analysis of the ribosomal RNAs can provide one of the best criteria to detect disease-induced, premature senescence. In work of this type a cautious approach must be used, since too heavy an infection, especially in an early stage of leaf development, is likely to result in an inhibition rather than a stimulation of RNA synthesis.

Ribosomes accumulate in tissues surrounding virus-induced local lesions. Lasting injury and the tissues surrounding virus-induced cells are necessary for such effects. If the damage is systemic or the toxic effect is too rapid, such changes do not take place or if they do, they are too fast to be observed. Thus, bacterial toxins can induce early senescence without previously (or locally) detectable synthetic phase

(Loviekovich *et al.,* 1964). High polysome/monosome rations are indicators of high synthetic rates. Stress conditions (drought, osmotic stress) are known to decrease the polysome/monosome ratio. It fits into the general picture decrease of chloroplast polysomes (Dyer and Scott, 1972).

Our knowledge of alterations in specific tRNAs by disease in higher plant cells is limited. This is surprising in view of the great interest in changes of tRNA is acceptor spectrum in relation to developmental processes, including senescence. In senescent tissues the decrease in the amount of total tRNA is generally less than that of the total rRNA/unit of DNA. This probably occurs in disease-induced senescence as well.

There has been much speculation on the role of mRNAs in host-parasite relationships. The limited evidence based on the appearance of DNA-like, rapidly labeled RNA exhibiting high turnover rate and template activity suggests that the increase in template-active mRNA synthesis might be one factor regulating the overall increase in synthetic activities (Tani *et al.,* 1973b). One recent work reports the isolation of poly(A) -containing mRNA from diseased tissues (Yoshikawa *et al.,* 1977).

The turnover and breakdown of nucleic acids have received less attention, although it is becoming increasingly evident that without determining these parameters it might be difficult to assess the synthetic rates. Thus, early work attributed a major, if not exclusive, role to increased RNA and protein synthesis when kinetin retards leaf senescence.

However, inhibition by kinetin of the breakdown of RNA and protein might be more important in maintaining steady RNA and protein levels. This important aspect of nucleic acid and protein metabolism has not been studied by plant pathologists. This is surprising because a possible connection between high synthetic activity in infection centers and local accumulation of cytokinins has repeatedly been suggested (El Hammady *et al.,* 1968).

In contrast to breakdown of RNA per se, much attention has been devoted to the study of increased nuclease levels, an ubiquitous phenomenon in disease and senescence. RNAases are known to be localized mainly in special cell compartments similar to the lysosomes of mamalian cells and the cellular RNAs are protected against them

(Matile, 1975). This explains that increase in RNAase level can occur concomitantly with an increase in total RNA. Therefore, the role of nucleases in host-parasite physiology remains to be elucidated. In the late stages of infection the breakdown of cellular organization might release the nucleases from their compartments and they can lead to a rapid degradation of nucleic acids, leading finally to cell death. However, it is imperative to emphasize that aging is an orderly process which cannot be based on a random degradation of macromolecules.

Protein Metabolism

The accumulation of ribosomes involves the synthesis of ribosomal protein and usually reflects a general stimulation of protein synthesis. Therefore, the pattern of protein synthesis in infected plants is similar to that already described for nucleic acids. In addition to the accumulation of ribosomes, the healthy cells surrounding the site of damage. The overall view is indicative of a synthetic metabolism geared to a higher level, supplied at the same time with the necessary energy-yielding machinery. The eliciting factors are not known. Probably no single wound hormone exist. One component of the *wound hormone complex* is ethylene which is invariably synthesized in injured cells and has been shown to regulate RNA and protein (enzyme) synthesis. However, ethylene stimulates the synthesis only of some enzymes and cannot be made responsible for the overall changes in protein level.

In contrast to nucleoytic enzymes, the proteases and peptidases have hardly been studied by plant pathologists. Although these enzymes have often implicated in aging, their role in protein breakdown and/or turnover is not known.

Abnormal Proteins

Aging in microbial and mammalian cells is often explained by the formation of abnormal proteins unable to carry out normal functions. The synthesis of such proteins could be due to somatic mutation, as well as to transcriptional or translational error. These changes are supposed to accumulate during ontogenesis and result in metabolic disturbances. The cell has built-in safety machinery to eliminate the abnormal proteins since they are more vulnerable to proteolytic attack because of their altered tertiary structure. All this is well documented for bacterial and animal cells. With plants one would be inclined to see a phylogenetic adaptation in the fact that, at least in some cases, proteolytic enzymes accumulate in leaf tissues as senescence proceeds. There appears to be the greatest need to eliminate aberrant proteins in this period. However, these bulk proteases are separated from the cellular sites where the safety machinery should work and probably play just the opposite role: they take part in the breakdown of total proteins at the onset of very advanced stages of senescence.

Our knowledge about proteases and their role in protein breakdown in plants is meager. The idea of aberrant proteins and a study of built-in defense mechanisms to eliminate them has not received any attention in relation to plant diseases and senescence, although such an approach might be highly rewarding.

Enzymic Patterns in Senescing Tissues

Since senescence is accelerated in detached leaves, this system has been widely used for studying aging phenomena. A whole set of enzyme activities is changed in detached leaves. Even if the different systems vary to some extent, on the basis of comparative studies we have put forward the idea that there is a *common pattern of enzymatic changes in detached leaves and tissues attacked by parasites* (Farkas *et al.*, 1964; Farkas and Lovrekovich, 1965). Ultranstructural studies have corroborated this view (Shaw and Manocha, 1965). It follows logically from this observation that a number of major enzyme changes in diseased tissues reflect the accelerated aging associated with disease.

The detached leaf system has some drawbacks. Bacterial contamination might induce enzyme changes attributed to detachment but no experimental evidence has been presented to support this view. A more serious problem is that leaf detachment or the use of leaf disks is invariably associated with mechanical damage, and injury-induced changes have to be distinguished from disease-induced changes. The drawback, however, can be turned into an advantage. Since senescence is not immediate the early changes are more likely to be due to cellular injury, whereas in the later phases senescence might predominate. The ubiquitous increase in RNAase activity in injured, senescent, and disease tissues might serve as a model example. In a number of disease situations a biphasic increase in RNAase activity has been observed. With rusts the two peaks seem to correspond to different enzymes (Chakavorty *et al.,* 1974). Logically, the enzyme responsible for the early rise could be an RNAase induced, by reaction of the host to shock or injury, and the other to the presence of large amounts of fungal cells or induced senescence. Indeed, we have shown that the overall increase in RNAase activity in excised leaves can be resolved into two components at least. The early rise is due to the accumulation of a relatively purine-specific RNAase, widespread in plant tissues (Wyen *et al.,* 1969), whereas in the later stages another, aging-specific nuclease accumlates, which exhibits relative adenine specificity (Wyen *et al.,* 1971a; Udvardy and Farkas, 1972). This process might well mimic the pathological situations, as suggested by the observation that in virus-infected local lesion hosts only the "injury-specific" nuclease increases under conditions when no senescence of the tissues is apparent (Wyen *et al.,* 1972). The term "injury" should be used in a broad sense, not restricted to mechanical damage, because the level of the same nuclease has been shown to increase dramatically in tobacco protoplasts exposed to osmotic stress (Lazar *et al.,* 1973; Premecz *et al.,* 1977).

Similarly, instructive is the investigation of peroxidases. Peroxidase activity also tends to increase in aging or injured or injured tissue as well as in diseased tissue. The total peroxidase activity can be resolved into several components. With peroxidases the chance is especially high of isolating artifacts (a point often overlooked by those who never tried to purify peroxidases and *separated* peroxidase isozymes by gel assay only). Therefore; reports on a very high number of peroxidase isozymes in crude extracts from infected tissues must be treated with reservation. Nevertheless, it seems possible to show, also in purified preparations, that senescence is associated with the accumulation of well-defined peroxidases. One such peroxidase appears prematurely in virus-infected young bean leaves (Farkas and Stahmann, 1966), an observation which gave experimental support to the common pattern hypothesis. However, in addition to aging, necrobiosis (lesion formation) per se has also a profound effect on peroxidases. In a number of host-virus combinations definite, predictable changes in peroxidase isozyme patterns were observed whenever lesions appeared. The changes depended on the host and not on the virus which induced the lesions (Solymosy *et al.,* 1967). This suggests that necrobiosis affects the formation of specific enzymes. Elegant studies on nearly isogenic, rust-infected wheats have also show that the necrogenic reaction is associated with the increase in one particular peroxidase isozyme, whereas the susceptible reaction favors the accumulation of a different one (Swvers *et al.,* 1971).

The level of peroxidase is quite stable once it has reached a peak. This is shown by the continued increase of peroxidase activity in the advanced stages of senescence, when the level of other "stress enzymes" starts to decline along with the excessive breakdown of total proteins (Lazar and Farkas, 1970). A further proof is the maintenance of the level of one peroxidase isozyme, specifically induced during resistant reaction, when the reaction type of the leaf, in which the resistant reaction had already developed, was shifted by experimental manipulation to a susceptible one, a reaction normally not associated with the formation of this particular isozyme (Seevers *et al.,* 1971).

Further typical stress enzymes which are generally present in injured present in injured and diseased tissues include the first two dehydrogenases of the hexose monophosphate shunt, enzymes involved in aromatic biosynthesis, and polyphenol oxidase. The reports are too numerous to list here.

As with nucleases and peroxidases, efforts have been made to see whether the elevated enzyme levels are associated with a preferential increase of specific molecular forms. Except for polyphenol oxidases in injured sweet potato (Hyodo and Uritani, 1967), the results are not very convincing.

The analysis of changed enzyme patterns is complicated by the greatly different turnover rates of different proteins. Thus, injury can induce the formation of enzyme-inactivating systems and decrease in enzyme level after an initial rise (Tanaka *et al.,* 1974). The stability of some other enzyme appears to be high,. Rubisco, e.g., under normal conditions, has hardly any turnover. However, it is broken down in disease and senescence.

Lysosomes

Since nucleic acids and proteins are quite stable, enzymes have to play a role in their breakdown. Specific processing by hydrolytic enzymes of nucleic acids and proteins is well documented for bacterial and animal cells. With plants the progress in this field is slow. One reason is the enzymes (nucleases, proteases, esterases) are contained in special cell compartments, lysosmes (Matile, 1975), and mostly sequestrated from plant metabolism. There is no evident reason why the plant needs such a large reservoir of hydrolytic enzymes that are seldom used during normal cell functioning. However, there must be some reason. We only know that the amount of lysosomal enzyme is greatly increased during aging and probably during disease. Perhaps, in plant disease and senescence, lysosomes temporarily play a protective role by compensating for a regulatory sysem that went wrong because of a defect in disease and/or senescence, and overproduced hydrolytic enzymes.

Secondary or Stress Metabolites

In the diseased tissues secondary or stress metabolites accumulate. With some oversimplification, two major pathways are responsible for the synthesis of following compounds:

- Aromatic biosynthesis
- Terpenoid biosynthesis.

Only the first has a major impact on senescence. Stimulated aromatic biosynthesis has long been known to be associated with necrobiosis. It is not clear why aging leads to a similar phenomenon. The difference, however, might be more phenomenological than basis. An essential attribute of senescence is the gradual breakdown of cellular structure. This is injury at the submicroscopic level. A plant organ does not have to be sliced or hervily infected to notice damage. The mostly forgotten observation that simple shaking of a leaf leads to respiratory increase can be interpreted as wound respiration due to systemic "mini" damage. It is even more remarkable that plants in outer space respond to the stress of weightlessness with an increased peroxidase activity.

The enzymatic machinery responsible for the accumulation of aromatic compounds in injured and/or infected storage tissues has been elucidated in detail. Such data, however, are almost completely lacking in relation to senescence.

Hormone Balance

The high frequency of hormone changes infected plants is odd, and is in contrast to the majority of animal and human diseases. The reason is in the peculiarities of hormonal regulation in the plant kingdom. On the one hand, several plant hormones simultaneously affect certain physiological phenomena. On the other hand, plant hormones are multitarget compounds, and one hormone can influence simultaneously (or sequentially) a large variety of different physiological processes. This network of hormonal regulation by a relatively low number of hormones explains that almost any change at any place in the network will lead to coupled changes in the interconnected processes.

Hormone changes in the infected tissues are too numerous to be listed here. We shall discuss only some cases which have a bearing on parasitically induced senescence.

Since infection often leads either to rejuvenation or to induced senescence, it is logical to explain these phenomena by alterations in the hormone levels. As already discussed, the large hormone pools (indoleacetic acid and cytokinin) accumulating in the infection centers of leaves attacked by biotrophic parasites are probably responsible for the transport of metabolites to the infection site, the formation of green islands, and the early senescence of the other parts of the leaf. Stimulation of ethylene formation is also a typical stress-induced process. The enhanced synthesis of ethylene induces fruit and leaf senescence, and a number of associated process. Perhaps the most remarkable is the dramatic increase in peroxidase level and the induction of phenylalanine ammonia lyase (PAL), an enzyme which might control the rate of aromatic biosynthesis. It has not been investigated whether the accumulation of phenolics during leaf senescence can be correlated with a more intense ethylene formation and a resulting increase in PAL activity. The available limited evidence suggests a rise and a later decline of PAL activity during seedling growth. However, PAL is a labile enzyme especially in the presence of phenolics and their oxidases. The detection of lower activities in old tissues does not necessarily mean lower PAL levels. This aspect of natural and disease-induced senescence is worth reinvestigating.

Increased Permeability

Diseased plant tissues, especially those reacting hypersensitively, exhibit increased permeability suggesting damage to the cell membranes. The ubiquitous increase in permeability in infected plants seems to be due to a general wound stimulus (Thilo and Nienhaus. 1975). The analysis of membrane lipid components after infection also reveals little specificity (Hoppe and Heitefuss, 1975).

Since ethylene is known to increase permeability, it might be a factor involved in permeability changes in senescence and disease. However, its mode of action is unknown and might be indirect. We only know that the permeability of cell membranes tends to increase with aging.

Temperature

Exposure to either high or low extremes of temperature can trigger leaf yellowing. The cytokinin level in the leaves also decreases considerably as a result of the heat shock. It is possible that the decline in cytokinin level might trigger the senescence. Growth of mature wheat plants at 35°c rather than 20°c results in marked decrease in their photosynthetic capacity. Although high temperatures initiate premature senescence, the degradation of certain thylakoid proteins including LHCPII is inhibited by the treatment (Ferguson *et al.,* 1993). This suggests that high temperature may disrupt some aspects of normal senescence. Thomas et al. (1980) have measured the rate of loss of chlorophyll from excised leaf sections of winter oat at a range of temperature from 5-20°c. They reported that the rate of senescence is reduced below 14°c. They concluded that a cold treatment may induce senescence but continued cold may slow the process once it has begun.

Water and Mineral Relations

It has been found that drought triggers leaf senescence and increases the rate of protein degradation. Premature senescence is also observed when plants are grown in waterlogged soil.

Leaf senescence is one of the typical symptom of mineral deficiency. If there is deficiency in mobile elements, such as nitrogen or phosphorus senescence of older leaves occur but younger leaves remain healthy because of nutrient remobilization from the older leaves. In contrast, if there is deficiency in a non-mobile element, such as calcium, only the young leaves are affected. It has been shown that nitrogen nutrition during leaf development greatly influences not only lifespan but the rate of senescence. A low level of nitrogen is correlated with a reduced cytokinin level in the shoots or xylem sap of several species. Thus, it is possible that the effect of nitrogen status on leaf senescence may be mediated by cytokinins (Singh *et al.,* 1992). When the supply of nitrogen is low, leaves of *Lolium perenne* with a high rate of respiration senesce more rapidly than those with a low rate of respiration (Pilbeam *et al.,* 1992), indicating that respiration may have a role in powering senescence.

Leaf senescence is unaffected by phosphorus deficiency. In general, plants appear to be remarkably efficient in their internal recycling of phosphorus. Under phosphorus starvation conditions, cell cultures of tomato and Arabidopsis have been shown to excrete an acid phosphatase. This enzyme increases phosphate availability by solubilizing organic phosphates in the rhizosphere. In addition, reduction of an extracellular ribonuclease has been detected on phosphate starvation of cultured tomato cells. Both of these enzymes have been proposed to be components of a phosphate-starvation-inducible rescue system in higher plants (Nurnberger *et al.,* 1990). Iron deficiency decreases the amount of photosynthetic pigments composition. Plants which are deficient in iron show a relative increase in xanthophylls, violaxanthin, antherasxanthin and zeaxanthin. The xanthophyll cycle functions in iron deficient plants but not in senescing leaves (Ababia *et al.,* 1991).

Necrotic lesions are also common symptom of iron deficiency or excess. Pennazio (1981) have noticed a similarity between these symptoms and hypersensitive response to virus infection in Gomphrena globasa. Three proteins localized in and around the lesions are of similar molecular weight to pathogenesis related proteins induced on virus infection while two others resemble proteins induced on leaf senescence. Azumi and Watanabe (1991) have cloned a senescence-enhanced cDNA from radish which shows homology to a pathogenesis related protein in tobacco.

Salinity and Senescence

Salinity conditions depress growth and photosynthesis in glycophytes or non halophytes. When the concentration of salt is higher, glycophytes begin show sign of growth inhibition, leaf discoloration and loss of weight. Evaporation and transpiration remove pure water (as vapor) from the soil, and this water loss concentrates solutes in soil. When irrigated water contains a high concentration of solutes and when there is no opportunity to flush out accumulated salts to drainage system, salts can quickly reach levels that are injurious to glycophytes. Dissolved solutes in the rooting zone generate a low osmotic potential that lowers the soil water potential. The general water balance of plant is thus affected, because leaves need to develop an even lower water potential to maintain a *downhill* gradient of water potential between the soil and leaves. Photosynthesis is inhibited when high concentration of Na^+ and/or Cl^- accumulate in chloroplast.

Chapter 7

FLOWERING AND SENESCENCE

Senescence can broadly be defined as the combination of events that leads to the death of cells, tissues or organs. There is tremendous diversity in the longevity of individual flowers. Some flowers persist for just a few hours, others last for about a day or several days and in the case of long-lived flowers, the flowers may persist for several weeks. The control of flower longevity is of interest to horticulturalists since cut flowers with a long vase life will have a greater commercial value, whilst in agriculture the length of the pollination period will influence fruit and seed set.

In exhaustive survey, Woltering and van Doorn (1988) studied the role of ethylene in petal senescence of flowers of 93 species representing 23 families. Except for a few families most of these flowers in which wilting was the primary senescence symptom were not sensitive to ethylene. Where petal abscission was the initial symptom of senescence, the flowers were mostly sensitive to exogenous ethylene. Maxie et al. (1973) demonstrated the close coincidence between the onset of the respiratory climacteric and a burst of ethylene synthesis in cut carnations. Woltering and Harren (1989) examined the role of ethylene in senescence of emasculated orchid flowers. The removal of pollinia in *Cymbidium* results in senescence of flower. The reason is that desiccation of the tissue of the rostrum stimulates the early production of trace amounts of ethylene. These minute quantities of ethylene are responsible for initiating general senescence of the flower.

ETHYLENE ACTION AND FLOWER SENESCENCE

The demonstration by Beyer (1976) that Ag^+ was an effective and specific inhibitor of ethylene action provided a powerful tool for examining the action of ethylene in plant growth and development. Using the more mobile silver thiosulphate (STS) complex, Veen (1978) demonstrated that inhibition of ethylene action greatly extends the life of cut carnations. This material has become an important commercial treatment for flowers sensitive to ethylene. The cyclic olefin Norbornadiene (NBD) also inhibits ethylene action by competitively inhibiting the ethylene binding site.

An hypothetical model for the action of ethylene in flower senescence has been proposed by Reid and Wu (1992). According to them the ethylene molecule binds to a site where the inhibitors of ethylene action, Ag^+ and NBD can also bind. When the binding site is sensitized and ethylene binds to it, a second message is generated which interacts with 5′ (promoter) regions of genes involved in ethylene – regulated senescence inducing transcription of the genes and synthesis of the proteins encoded by the genes on polyribosome. A considerable amount of evidence has accumulated from studies with cut flowers that is consistent with this scheme.

STRUCTURAL CHANGES OCCURRING DURING FLORAL SENESCENCE

In cut flowers ethylene accelerates petal wilting while the natural senescence is accompanied by increased ethylene production. In carnation (Dianthus caryophyllous) the senescence of flowers is accompanied by the inrolling of the petal margins. This process is accelerated by ethylene and pollination. Within 3h of pollination the production of ethylene from the stigmas increases 10-fold at which time a small increase in ethylene production from the ovary and receptacle is detectable (Nichols *et al.,* 1983). The increase ethylene production suggests that this is the result of an interaction between the growing pollen tubes and the stigmatic tissues. In Petunia hybrida pollination brings about the collapse of the corolla. The growing pollen tubes bring about a wounding like response and elicit rapid ethylene production (Gilissen and Hoekstra, 1984). It has been demonstrated that as a result of pollination a water soluble substance is translocated down the style. Whitenead et al. (1983) suggested that ACC may be translocated between the flower parts. Other suggestion is that the short chain fatty acids such as octanoic and decanoic acid increase after pollination and application of small droplets of these chemicals to the stigma increase the sensitivity of the flowers to ethylene. Lovell et al. (1987) suggested that the intact stigma may produce a compound which prevents senescence and that it is the removal of the source of this compound by either wounding or pollination that bring about accelerated wilting.

COLOUR CHANGE OCCURRING DURING FLORAL SENESCENCE

Perianth colour changes occur in large number of angiosperm families. In *Lupinus albifrons,* the colour change is accelerated by either pollination or treatment with ethylene. Once the colour charge is complete, the production of ethylene decreases significantly. Burg and Dijkman (1967) have reported two peaks of ethylene production. Woltering (1988) shown that the first peak is associated with anthocyanins accumulation and the later peak with the wilting of the petals and sepals.

GENE EXPRESSION DURING FLORAL SENESCENCE

Woodson and his group (1989) have provided evidence that aging of carnation flowers and the onset of ethylene biosynthesis are accompanied by synthesis of mRNA species not found in young flowers. There are two classes of mRNAs that increase in abundance during petal senescence. One mRNA (pSR5) is present at low levels during the early stages of development and begins to accumulate in mature petal prior to the increase is ethylene production. Accumulation of this mRNA is retarded by inhibiting the synthesis or action of ethylene. The other class of mRNAs are highly regulated by ethylene. The other class of mRNAs are highly regulated by ethylene. They are not detectable prior to the rise in ethylene production an dincrease in abundance with ethylene climacteri. Their expression is markedly inhibited by inhibition of ethylene production or action. The rapid rise in polyribosome profiles shows that new genes are transcribed during this period. The treatment of flower petals with inhibitors of protein synthesis substantially delays their senescence. Increased activities in catabolic enzymes are associated with decline in macromolecules including DNA, RNA, proteins and membrane lipids. The senescence of petals is a careful regulated event involving the synthesis of specific proteins.

REGULATION OF FLOWER SENESCENCE BY PLANT HORMONES

Members of each of the five groups of plant hormones have been implicated in the regulation of senescence. It is generally accepted, that control is exerted through a balance between plant hormones interacting with each other and with other internal factors. However, ethylene seems to have been dealt with most extensively.

Ethylene

A wide range of flowers are, to various degrees such as rolling of petals, fading of petal, wilting of petal, affected by ethylene, with few responding with typical symptoms.

A wide assortment of flowers producing ethylene (Fischer, 1950; Frenkel *et al.,* 1976) also seem to posses the capacity to metabolize various substrates into ethylene (Phan *et al.,* 1970). In many studies, ethylene emanation by vegetative or flower tissue was thought to reflect changes in ethylene production. It has been pointed out that this may not necessarily be so, because the emanation might also be influenced by changes either in the capacity of sites of ethylene attachment (Jerie *et al.,* 1979) or ethylene metabolism rates. Although this hypothesis helps to explain the possible involvement of ethylene in the regulation of senescence processes.

The time course of ethylene production follows a typical profile composed of 3 distinct phase: first, a low steady rate of ethylene production, second, an accelerated rise of ethylene production to a peak, followed by a last phase. in which ethylene production is declining. Various events associated with senescence can be studied with reference to these phases. Usually the visual symptoms occur only in the second phase, and are pronounced toward the end of this phase. Thus, the onset of this phase is of importance, because it signals the terminal step of senescence.

Application of CO_2 to carnations during the first phase of ethylene production will delay the terminal phase of senescence. However, application of CO_2 at the start of the onset of the second phase is sometimes ineffective in delaying the transition into the auto stimulation phase of ethylene production. The onset of the second phase may be induced by various means, such as exposure to ethylene (Nichols, 1968) pollination (Hall *et al.,* 1967; Burg *et al.,*

1967; Nichols, 1977) and abscisic acid (Mayak *et al.*, 1976; Nichols, 1977). By contrast, it can be delayed by CO_2, (Mayak *et al.*, 1976; Nichols, 1977; Smith *et al.*, 1966; Smith *et al.*, 1966) low concentration of oxygen (Nichols, 1968) and hydobaric pressure environment (Mayak *et al.*, 1976;. Nichols, 1977).

Kende and Baumgartner (1974) have proposed a model based on changes in compartmentation to account for ethylene-induced ethylene synthesis: by increasing the permeability of the tonoplast, ethylene could enhance the flow of substrates from the vacuole to the cytoplasm, where an ethylene-generating system is located. This model is supported by experiments in which flower tissue was preloaded with radioactive tracers. representing either the vacuole or cytoplasm content, then exposed to ethylene and subjected to compartmental analysis (Macrobbie, 1971).

The presence of separate storage and metabolic amino acid pools has been suggested in higher plants (Oaks *et al.*, 1970). Although clearly demonstrated in yeast (Wiemken Gohring *et al.*, 1974) the vacuoles of higher plants probably serve as a storage place for amino acids including methionine.

Numerous studies on the biosynthesis of ethylene have been conduced, in which several ethylene-forming systems were evaluated. Methionine (Adams *et al.*, 1977) is the only generally accepted precursor of ethylene in the tissue of higher plants (Varner *et al.*, 1976). The existence of separate amino acid pools suggests that the tonoplast membrane may act as rate-limiting control in the synthesis of ethylene, thus indirectly lending support to the above proposed model.

Exposure of mature carnation flower tissue to ethylene will generally hasten senescence with the development of typical symptoms. These symptoms will last even when the ethylene is ventilated away, are irreversible (Mayak *et al.*, 1976) and are presumably associated with a rise in ethylene synthesis. Although the young flower will develop the same symptoms upon exposure to ethylene. If the ethylene is ventilated away, even though it had been present to the point when symptoms become visible, the flowers will recover and resume their natural course of development. Hence, the tissues vary in their response to the presence of ethylene. Young flowers are not responsive, whereas mature flowers respond - and transition into the second phase occurs. Perhaps while ethylene content in flower tissue is not markedly augmented until the senescence phase, enough gas is present during the presence phase. The auto-stimulation of ethylene production occurring in senescence will be initiated once the sensitivity to ethylene has changed to respond to the existing ethylene.

The sensitivity to ethylene can be the result of an intricate, complex interaction between internal factors such as plant hormones (Sacher, 1973) like abscisic acid (Nichols, 1977) carbohydrate reserve (Mayak *et al.,* 1975) and osmotic concentration of the petal tissue (Mayak *et al.,* 1978). The flower is composed of various organs: petals, sepal, style, ovary, etc., each at a different phase of its development. The compounded result of processes and internal interactions, occurring at each phase of development, may effect the response otherwise referred to as sensitivity to ethylene. The responding systems in each organ gradually become operative at different rates and times as the flower organs mature.

The entire process of ethylene-induced senescence may be under programmed regulation or control. However, the responding system may be affected by environmental factors. Temperature (Maxie *et al.,* 1973; Smith *et al.,* 1966) water stress to cut flowers imposed by excessive drying, and cold storage used in the practice of handling flowers (Mayak et al, 1976) all result in higher sensitivity to ethylene.,

Generally, the responding system gradually becomes more sensitive as the flower organs mature. The petals may respond to ethylene generated in the style, which was demonstrated to have a high capacity for ethylene synthesis (Burg *et al.,* 1967; Hall *et al.,* 1967; Lipe *et al.,* 1973; Burg and Dijkman, Burg *et al.,* 1967) using *Vanda* orchids in their studies, proposed that following pollination. a transfer of auxin from the pollen to the stigma takes place and is spread through the column.

The growth hormone triggers a chain reaction of ethylene synthesis in the cells so that ethylene becomes present throughout the flower, with the net result of petal fading. A similar proposition was put forward to explain the wilting of carnations occurring after pollination (Nichols, 1977). Here the style, through its endogenous ethylene production, is thought to trigger petal wilting. Gillisen (1977) using *Petunia* flowers, reported that the penetrating pollen tubes activate the stylar tissue, which then leads to the wilting of the flower. The rate of wilting depends on the number of viable pollen grains penetrating the stigma.

In conjunction with the petal wilting, swelling of ovary occurs (Nichols, 1968). Nichols (1976) also found that ethylene promotes an accumulation of sugars and inorganic materials in the ovary, accompanied by loss of fresh and dry weight of the petals. The swelling of the ovary appears to result from the enlargement of existing cells. Ethylene assumes the role of a synchronizer: by induction of total wilting it facilitates substantial mobilization of carbohydrates from the petal, providing for the development of the ovary, The hormonal regulation of a physiological process may be accompanied by a proportionality mechanism (Varner *et al.,* 1976). Thus, a process is enhanced as concentration increase, and diminishes as the concentration drops. Accordingly, the changes in plant hormone concentrations may be interpreted as being signals controlling the processes. Indeed, the tissue response to ethylene is proportional to the log of the concentration of ethylene present. The level of ethylene in the tissue, as it changes with time, as well as being part of a controlling signal, is determined by the rate of biosynthesis and its diffusion along a concentration gradient into the surrounding tissue and into the atmosphere.

Cytokinin, Auxin and Gibberellins

The endogenous level of cytokinins in rose petals decreases as the flower ages, and the level is lower in a short-lived cultivar than in a long-lived one (Mayak *et al.,* 1970). Results with external application of cytokinins, which delay senescence of various flowers (Mayak *et al.,* 1976; Mayak *et al.,* 1970) support the possibility that the reduction of the internal level of this growth regulator is associated with senescence processes in plants. Further support for this supposition may be drawn from experiments (Mayak *et al.,* 1974) in which applied kinetin slowed down the reduction in dry weight that took place as the flower matured. In addition, water uptake was enhanced, and it was postulated that this was achieved by maintenance of cell integrity. However, the effects of kinetin and other cytokinins on the senescence of leaves are more complex.

The presence of gibberellins and auxins has been reported in carnations (Jeffcoat, 1977). Gibberellic acid increases flower size and extends the longevity of carnations (Kelley *et al.,* 1964; Garrid *et al.,*

1978). Gilbart and Sink (1971) assigned to auxin a central role in the control of senescence in poinsettia.

The auxin level decreased in two poinsettia cultivars, but faster in the short-lived cultivar. Also the activity of IAA-oxidase and the level of hydrogen peroxide increased with aging These researchers hypothesized that auxin delays senescence by inducing the synthesis of peroxidase, which then prevents the accumulation of free peroxide associated with aging. Auxin may be linked to factors determining sensitivity to ethylene discussed earlier. However, further studies are needed with various flowers to substantiate the involvement of these plant hormones with senescence.

Abscisic Acid

The role attributed to abscisic acid in petals is that of regulating the process of senescence. The endogenous level of abscisic acid rises as petals senesce. It rises to a higher level in a short-lived variety than in a long-lived one.(Mayak *et al.,* 1972). The exogenous application of ABA accelerated senescence of carnations (Nichols, 1977) as well as that of roses.

Kende and Hanson (1976) reported that benzyladenine delayed ethylene production by isolated rib segments of *Ipomoea* flowers. Similar results were reported with carnation flowers, and their responsiveness was also affected (Eisinger, 1977) a plant hormone may thus affect the level of another plant hormone, thereby having more intensive effect on the hormonal balance.

Transition into the phase of ethylene auto-stimulation may occur at an earlier stage. The permeability of the tonoplast (Hanson *et al.,* 1975; Mayak *et al.,* 1977) increases. Leading to cell dis-organiment accompanied by a reduction in water uptake and plausible development of water or ion stress. The well established stress response. i.e., increase in the level of abscisic acid (Livne *et al.,* 1972) soon follows. In this respect, the rise in the level of ABA. in response to ethylene as observed with cut rose flowers .94 can be interpreted as a secondary result via an effect on membrane permeability and water stress imposed by ethylene.

Events reflecting the effect of ethylene on the level of abscisic acid are mediated by membrane modifications. It will be interest to study whether or not the reciprocal effects of cytokinin and ABA on each other's activity, suggested to be a widely occurring phenomenon in plants by Itai and Benzioni, (1976) is also mėdiated by membrane modifications. The observation that kinetin delays wilting of cut rose flowers by protecting cell integrity (Mayak and Halevy, 1974) supports this possibility. It is conceivable that if cell integrity is maintained, the development of a situation of stress is delayed and abscisic acid will be kept at a low level. along with the increase in membrane permeability, a reduction in phospholipid content is affected.(Beutelmann *et al.,* 1977). A reduction in phospholipid was found in carnations (Ronen, 1978) also in the presence of rhizobitoxine, a specific inhibitor of the conversion of methionine to ethylene (Owens *et al.,* 1971). Hence, the reduction in phospholipid in senescent carnation petals is not necessarily dependent on methionine-dependent ethylene synthesis.

Chapter 8

SENESCENCE IN SEEDS

Each seed contains a potential new individual of its life and it is adapted to wide range of unfavorable conditions for growth. Stored dry deed survives longer and undergoes in senescence slowly comparatively than imbibed seed. It means there is difference between senescence pattern of dry and imbibed seeds. In dry seeds rate of senescence reduces due to metabolically inactivity.

SENESCENCE CHANGES IN DRY SEEDS

Changes Associated with Membranes

Electron micrographs of cells of non-viable (or dead) embryos of rye (Hallam et al, 1973) show that the outer membrane of many mitochondria is extended to form a "blister" and is dissociated from a broken inner membrane. The plasma membrane is frequently contracted from the cell wall, and the presence of electron-dense material in the area between the two indicates that the semipermeability and molecular integrity of the plasma membrane have been disrupted. The nuclear membrane, too, may show irregularities and breaks not present in cells of a viable embryo. Although some of the membrane changes seen in a dry non-viable embryo may the exacerbated by aqueous fixation procedures, they clearly reflect a difference existing between the membranes present in viable and non-viable cells.

The progressive loss of integrity of the plasma membrane is readily demonstrated in aging seed by the extent of leakage of cytoplasmic components to the external medium when the dry seed are put into water (Harman *et al.,* 1972; Parrish *et al.,* 1978). All seeds show some loss of solutes as measured by the rise in conductivity of the external solution, but in germinating embryos of high vigor, sugars, amino acids, and other metabolites are subsequently reabsorbed by active uptake. Depending upon the increasing degree of senescence of the embryo and progressive impairment of the membranes, the initial loss of cytoplasmic solutes is correspondingly greater, and the extent of subsequent uptake less. One of the events in a germinating embryo, therefore, is the apparent restoration of the semipermeable property of the plasma membrane and repair of lesions in membrane integrity that occurred when the seed was dry. That membrane synthesis and repair does occur at imbibition is apparent from ultrastructural studies. In maize, for example, lysosomal engulfment of damaged organelles, (e.g., mitochondria) and their subsequent renewal follows closely upon imbibition (Villiers, 1971; Berjak *et al.,* 1970; Berjak *et al.,* 1972).

It has been proposed that loss of membrane integrity in the dry state could involve the formation of free radicals and the peroxidation of unsaturated fatty acid components. Indeed, the detrimental involvement of free radicals could be inferred from the reported decrease in unsaturated fatty acids during aging in embryonic axes of pea seeds. The levels of trienoic acids, (e.g., linolenic) decreasing more rapidly than those of dienoic acids, (e.g., linoleic) (Harman *et al.,* 1976) A declining ability of a germination seedling to repair the membrane lesions that occur in dry seeds would be a major factor leading to poor germination.

The decreased respiratory activity of aged seed at imbibition must have important consequences for limiting synthetic processes. and can be correlated with the poor integrity of mitochondrial membranes observed in electron micrographs of sections of *aged* or non-viable material (Hallam *et al.,* 1973).

Electron micrographs of the pellets of mitochondria showed that those from the older seed source possessed swollen cristae and lacked their outer membranes. Such results are further evidence that age-related deleterious membrane changes do occur in mitochondria during seed senescence.

Effect on Germination ability

It is found that newly harvested seeds germinate with speed, and respiratory activity (Abdul-Baki, 1969), ATP generation (Obendorf,

1974), protein synthesis (Marcus *et al.,* 1966), and all classes of RNA synthesis (Sen *et al.,* 1975) can commence within minutes.

Seeds that have been stored for long periods of time (even in favorable conditions) germinate more slowly than fresh seed. and initially show low rates of protein and RNA synthesis (Sen and Osborne, 1974). In addition, the critical time at which DNA replication is initiated becomes progressively delayed (Sen and Osborne, 1974). The old seed, though viable, expresses all the symptoms of loss of vigor. The senescent changes are sufficiently extensive, the rates of synthesis of new cytoplasmic components remain at such a low level that the embryo fails to establish as seedlings. It seems that cells of a seed are equally susceptible to senescent change.

In barley, aleurone cells exhibit impaired hydrolytic α-amylase activity with reduced germination performance of the embryo. Neither the production of α-amylase, nor germination, is improved by the addition of gibberellin (Aspinall and Paleg, 1971) indicating that a low level of hormone does not limit the performance. Further evidence that each part of the seed deteriorates on storage comes from embryo transplant experiments in wheat (Floris, 1970). When embryo of low vigor were attached to aleurone parts of high vigor (or the reverse transplant made), aged material was not improved and vigorous was impaired.

The greatest impairment of protein synthesis (or the greatest rate of senescence change) occurs very early in the life span of the dry embryo, during the loss of viability from 95% to 86%. In embryos isolated from seed that has just reached the condition at which it cannot germinate (0%).

Respiratory Changes and ATP Formation

ATP generation is reduced due to failure of mitochondrial function. A general event in aging seeds is the progressive loss of activity of mitochondrial dehydrogenases with seed vigor (Throneberry and Smith, 1955) culminating in the complete loss of activity at the stage at which the embryo can no longer synthesize protein or germinate. Failure of dehydrogenase function is the basis of an *in situ* test for viability in the early hours of imbibition (Lakon, 1949) whereas living cells will reduce a solution of the colorless 2,3,5-triphenyltetrazolium chloride to the stable, red-colored non-diffusible formazan within the tissues, the dead cells remain colorless.

In high vigor embroyos, the activity of high energy nucleoside triphosphate generating systems and the rise in ATP levels during the first 20 min of germination can be dramatic (Obendorf and Marcus, 1974). It is not surprising, therefore, that at imbibition any loss of mitochondrial function should be associated with low levels of ATP production (Ching,1973) and reduced rates of ATP and GTP-dependent events in protein synthesis (Roberts and Osborne, 1973).

Changes Related to Protein Synthesis

Neither new messenger RNA nor new ribosomes are synthesized, and poses the question of whether (in a senescent seed) this is due to failures in the mechanism of transcription from the genome, an extensive and rapid degradation of RNA by nucleases, or an impairment of the nuclear DNA template.

The concept that messenger RNA synthesized during the development of the seed should remain stored within the cells during the period of dehydration and quiescence, and that the same messenger RNA should at once become available to direct protein synthesis when water is again provided at germination, has considerable attraction and theoretically permits the commencement of protein synthesis in the absence of new mRNA synthesis (Marcus and Feeley, 1964; Dure and Waters, 1965).

Using incorporation of labeled amino acid precursors into acid - insoluble products as a measure of protein synthesis in cell-free system, it has been shown that fractions containing mRNA template activity can be isolated from a variety of different dry seeds, including wheat (Chen *et al.*, 1968) rice (Bhat and Padayatty, 1974). However, amino acid incorporation into unidentified acid-insoluble products in a cell-free system, or into products separable into discrete fractions by electrophoresis in polyacrylamide gels, does not provide unequivocal evidence for a similar messenger effectiveness for translation into defined proteins in vivo. Nor does it indicate whether the proteins, if formed, are important to the reactivation of synthetic activity at embryo germination.

On the basis of experiments with cell-free systems prepared from dry embryos of wheat (Weeks and Marcus, 1971; Spiegel and Marcus, 1975). it was proposed that the early protein synthesis at germination could be coded for by long-lived messenger RNAs and be translated on old ribosomes that are stable to storage in the dry seed. Evidence suggests an alternative; namely, that the early protein synthesis could be coded for by newly synthesized messenger RNA molecules transcribed at imbibition (Sen and Osborne, 1977; Dobrzanska *et al.*, 1973). Because the assembly of newly synthesized ribosomes and their transfer to the cytoplasm apparently takes longer than the activation of protein synthesis, it is clear that old ribosomes can be the sites of early protein synthesis (Sen amd Osborne, 1977).

That lesions do occur both in ribosomes and in components of the post-ribosomal supernatant fraction containing tRNA, aminoacyl-tRNA synthetases, and transferase enzymes was first demonstrated from studies with cell-free systems (Roberts and Osborne, 1973; Roberts and Osborne, 1972; Szezotka, 1975) isolated from dry rye embryos of different percentage viability.

The progressive loss of integrity of ribosomal RNA in dry seed has been demonstrated during senescence in dry rye embroys. Extraction of total RNA or of rRNA from ribosomes from a range of embryos of different percentage viability, followed by fractionation of the high molecular weight RNA components by electrophoresis of polyacrylamide gels, has shown that 18S rRNA molecules suffer increasing degradation with loss of viability, yielding fragments appearing to be of defined molecular weight (Roberts et al, 1973). A progressively increased number of breaks with decreasing viability were revealed in 25S rRNA by disrupting the hydrogen bonding in helical regions of the RNA, (e.g., by heating a solution of the RNA, followed by rapid cooling) before fractionation by electrophoresis on polyacrylamide gels. Although most of the ribosomes from 98% viable embryos contain undegraded RNA, a proportion of the embryos having lower viabilities contain ribosomes with fragmented 25S and 18S rRNA, the proporation of impaired ribosomes being higher in the lower viability stocks (Osborne et al, 1974; Roberts and Osborne, 1973). There is, however, no evidence of a loss of components present in the protein-RNA complex of the ribosome, for all ribosomes sediment at the same rate in sucrose density gradients (Roberts et al, 1973). Since the rRNA obtained by direct extraction from the dry embryo both rRNA obtained by extraction from ribosomes first isolated from the dry seed.

Ribonucleases, unlike dehydrogenases (Throneberry and Smith, 1955) are highly stable during the extended storage of dry seed. In authentic 103-year-old wheat (which must have been non-viable for at least 60 years) active ribonucleases could still be extracted (Osborne et al, 1974). During and after loss of viability in a seed it seems that a continuous (but low) rate of nuclease mediated degradation of ribosomal RNA to small molecular fragments or nucleotides.

The separation and assessment of the activity of post-ribosomal supernatant factors essential for active peptide bond formation has been carried out in cell-free systems with preparations from seeds of different percentage viability. Some components retain full activity at loss of viability, while others are impaired. For example, in rye, phenylalanyl-tRNA and phenylalanyl-tRNA synthetase appear to be equally active in sustaining polyphenylalanine synthesis in preparations from embryos of any of the viabilities tested in homologous or heterologous systems coded for by Poly U. Failure of polypeptide synthesis in cell-free systems prepared from rye of low viability must therefore occur at some stage coincident with, or subsequent to, the binding of aminoacyl-tRNA to the ribosome (Roberts and Osborne, 1973). Species may differ in the relative degree of integrity of these components of the protein synthetase has been reported for aged pea seeds (Bray and Chow, 1976).

Studies of the transfer enzymes in rye, pea, and wheat all indicate that the translocase or elongation factor as (Transferase enzyme II) concerned with transfer of the amino acid from the aminoacyl to the peptidyl site on the ribosome show only slight impairment with loss of viability. However, the GTP-dependent binding of aminoacylated phenylalanyl-tRNA to ribosome, determined by Transferase I, decreased in proportion to the loss of activity in the homologous system for polyphenylalanine synthesis (Roberts and Osborne, 1973). The fall in Trasferase I activity also parallels the loss of protein synthesizing acitvity in embryos (Roberts and Osborne, 1973) or seeds (Bray and Chow, 1976; Dell'Aquilla *et al.,* 1976) *in vivo* and closely reflects the extent of cellular senescence in the seed.

With loss of viability and the general decline in capacity for protein synthesis in the early hours of germination, there is a parallel, but depression in the rates of synthesis of all classes of RNA (Sen and Osborne, 1977). In rye this is accompanied by a reduced rate of processing of precursor ribosomal RNA (31S) to 25S and 18S components (Osboren et al, 1974; Sen and Osborne, 1977) and at 0% viability by a synthesis of RNA molecules no greater in molecular weight than 4 to 5S (Osborne, 1977). The synthesis and degradation of this RNA and its failure to be utilized in protein synthesis by 0% viable material (Sen and Osborne, 1977) could be due to a variety of causes. One possibility is that this is a nonsense RNA transcribed from an impaired template. Given that a transcriptional mechanism is operative with functional polymerases, it is of particular interest to consider the integrity and biochemical status of DNA during seed senescence.

The total level of DNA present in the nucleus of cells of rye embryos remains constant during senescence, and for at least a year after loss of viability (Roberts and Osborne, 1973). The molecular integrity of the DNA does not, however, remain unchanged during this period, for the amount of high molecular weight material that can be recovered by spooling techniques decreases with loss of viability (10.2 mg of DNA per gram in 90% viable dry embryos compared with 3.1 mg of DNA per gram of non-viable embryos (Roberts and Osborne, 1973). Breaks in DNA could contribute to the much reduced template activity, and possibly also to faulty transcription. Breaks might also be the cause of the absence of high molecular weight RNA products at transcription: the experiments with 0% viable rye embryos, which do not synthesize protein, but do synthesize low molecular weight (4-5S) polyribonucleotides located in the nucleus do imply that the DNA template may be highly fragmented, though it cannot be excluded that the polynucleotides represent fragments of rapidly degraded high molecular weight transcripts (Sen and Osborne, 1977; Osborne *et al.*, 1977).

If newly synthesized RNA is required for early protein synthesis, then any failure in the DNA template or loss in fidelity of transcription must lead to eventual cell death. Certainly, the products transcribed by 0% viable rye embryos (Sen and Osborne, 1977; Osborne *et al.*, 1977) or pea embryonic axis (Payne *et al.*, 1977) do not code for protein synthesis *in vivo* and are associated with cell death.

De Vries in 1901 first questioned that changes in genetic material might be a cause of poor germination, and account for the high proportion of abnormal plants produced from old seed. Not until the 1930s experiments were planned to determine whether the extent of mutation was related to seed age. The correlation then became very clear. There was a greater frequency of cytologically aberrant mitosis in root tips on germination with increasing seed age (Roberts, 1972;

Nilsson, 1931; Navashin, 1933); and a higher number of chromosomally aberrant pollen grains were produced by plants from aged or accelerated aged seeds (Jenns and Bryant, 1978).

Chromosomal aberrations probably give a low estimate of the extent of genetic damage, as they are visible only when the cells divide at mitosis or during meiosis in pollen formation. It is interesting, therefore, that seeds stored under nitrogen show fewer chromosomal disturbances (Harrison, 1966) while increased oxygen concentration increased the numbers of aberrant chromosomes (Roberts and Abdalla, 1968; Abdalla and Roberts, 1968). It was not clear, however, whether damage to DNA occurred in situ during senescence of the dry seed, or was an early event that followed imbibition.

Elctrophoretic separation of the DNA from nuclei of non-viable and 95% viable embryos, which is lysed directly onto polyacrylamide gels, showed that the DNA of 95% nuclei migrates only a short distance into the gels, but DNA from the nuclei of non-viable material possesses an additional disperse DNA component that spreads throughout the gel, and is of a higher molecular weight than the nucleosome monomer.

Since total levels of DNA do not decline in the nuclei of embryos during senescence (Roberts and Osborne, 1973) the most likely cause of loss of DNA integrity is the in situ activity of endodeoxyribonucleases, possible with some additional degradation due to exonuclease activity. If such breaks resulting from endonuclease action occur before the embryos become 0% viable, then they could contribute to a progressive failure of the DNA template to code for complete RNA transcripts at germination.

Tests for DNAases in rye embryos showed their activity to be higher in the non-viable than the 95% viable dry material ((Roberts and Osborne, 1973) Since it is most unlikely that enzyme synthesis occurs in the dry embryo, this difference must result either from some form of activation of pre-existing latent enzyme, or from a preferential stability or extractability of nuclease from non-viable material. Experiments suggest that preparations of 95% viable embryos contain a heat-labile inhibitor of DNAase that may become inactivated with loss of viability (Cheah and Osborne, 1978). If the endodeoxyribonucleases are associated with the nuclear chromatin, even in seeds with moisture contents of 10% or less, slow enzymic DNA fragmentation could occur.

There is limited evidence for DNA repair systems in plants. In root tips (of germinating onions), the number of chromosomal aberrations was found to decline following repeated cell divisions (Nichols, 1942) suggesting either a failure of divisions in cells with imparied nuclei, or a possible DNA repair process in the newly imbibed cells (Viller, 1971; Berjak and Villers, 1970; Berjak and Villers, 1972).

In viable rye embroys increasing levels of non-replicative incorporation ^{3}H-Methymidine into DNA is an early event (within 60 min) in germination in response to increasing doses of γ-irradiation to them in the dry state. This indicates a potential for repair of single-stranded DNA breaks. Repair of chemically-induced single-stranded breaks has been shown also in barley embryos (Veleminsky et al, 1973; Soyfer *et al.,* 1977). The repair of irradiation-induced chromosomal disturbances may have been demonstrated in imbibed but in dormant lettuce seed. Repair of lesins seems unlikely, however, during dry storage of seeds, and an accumulation of breaks in DNA could eventually reach a level too high to be repaired at germination. Furthermore, if the DNA could eventually reach a level too high to be repaired at germination. Furthermore, if the DNA repair system were composed of some of the more labile proteins of the cell., then in embryos of low viability, repair could be defective or completely lacking-with little chance of the appropriate enzymes being synthesized a new. It may be pertinent, therefore, that in most germinating seeds, the on set of DNA replication before cell division becomes progressively delayed, the lower the viability of the material (Sen and Osborne, 1974). This could be a reflection of deficiencies in repairing the DNA template prior to DNA replication and of the loss of DNA polymerase activity as reported for ageing embryos of wheat (Soyfer *et al.,* 1977) and in nuclei of 5-year-old barley seeds.

In pea, a rise in deoxyribonuclease activity is associated with the initiation of DNA replication (Jenns and Bryant, 1978) and may be concerned with the formation (by endonuclease cleavage) of initiation sites for synthesis of limited length transcripts of DNA. An inhibitor of this DNAase activity must be present in viable seeds and had been demonstrated in imbibed material up to the time of the replication event (Jenns and Bryant, 1978). Further, evidence that a DNAase is involved in early DNA replication comes from work with wheat (Buchowicz et al, 1978) here, all the early DNA synthesized prior to replication was of low molecular weight, some presumable being transcribed from low molecular weight DNA templates. It could be envisaged that the presence of a stable, endodeoxyribonuclease in the dry seed, together with a labile inhibitor of the enzyme, the presence of a stable endodeoxyribonuclease in the dry seed. together with a labile inhibitor of the enzyme, could result in viable seed with suppressed endonuclease until the time of DNA replication at germination. During seed senescence, the nuclease inhibitor could undergo slow inactivation so that eventually the endonuclease would be activated before slow inactivation so that eventually the endonuclease would be activated before germination. DNA could then suffer fragmentation and degradation in the dry seed. If the DNA repair enzymes are as labile as the inhibitor of DNAase then, the eventual irreparable loss of DNA integrity is a program for cell death.

CHANGES IN IMBIBED SEEDS

The only seeds that do not grow when placed in water are those that are senescent or dormant. When a dead seed or a 0° viable seed (i.e., one incapable of protein synthesis) is placed in water, the relatively stable hydrolytic enzymes present in the cells continue to function and fragmentation of macromolecules and cytoplasmic disorganization proceeds progressively. However, if a dormant seed is placed in water all the normal metabolic processes are reactivated with the exception of those directed to cell growth and DNA replication. A dormant embryo can sustain this imbibed state of arrested development for months or years with the normal biochemical events, e.g., respiratory activity (Simmonds and Simpson, 1972) protein synthesis (Cuming and Osborne, 1978; Chen and Varner, 1970; Hecker and Berhardt, 1976) RNA synthesis (Jarvis *et al.,* 1968) and lipid metabolism (Bradbeer and Colman, 1967) in continuous operation. An active polysome formation (Fountain and Bewley, 1973) the synthesis of polyA-rich RNA (Hecker, 1976) and a constant turnover of membrane protein (Cuming and Osborne, 1978) and phospholipid (Cuming and Osborne, 1978) are characteristic of the dormant imbibed condition.

In dormant imbibed state if the metabolic turnover is such that a true homeostatic condition obtains, then there can be no accumulation of lesions or loss of integrity of cytoplasmic organization, for constant turnover should maintain an error-free system. No specific repair processes need be invoked to account for the maintenance of full cellular function. With all the evidence for an active metabolic turnover of the normal maintenance processes that ensure cellular integrity, a progression of senescence events is difficult to envisage. Why then, imbibed seeds fail to survive indefinitely?

One possibility is that storage reserves become depleted, although there is no evidence that this is so. Another possibility is that in the absence of other degenerative changes in the cells, the lethal accumulation of DNA lesions might take much longer to occur than in those of a dry seed. Indeed, experiments with dormant lettuce (Aksenov, 1969) have shown that seeds stored dry for various lengths of time show increasing numbers of chromosomal aberrations at germination, whereas those stored full imbibed show essentially none (Villers, 1974). Furthermore, when dry seeds showing different levels of chromosomal aberrations were then imbibed (but maintained under non-germinating conditions), they showed no further increase in aberrations with time. However, although the loss of DNA integrity was halted, there was no significant evidence (as judged by the number of aberrant cells at subsequent germination) of a repair of the DNA lesions.

The suppression of DNA synthesis may not be complete. however, for although by incorporation studies a lack of DNA replication can be verified, it is difficult to exclude the possibility of some DNA repair synthesis (Hecker and Huth, 1975). Other experiments in which dormant lettuce seeds have been exposed to gamma-irradiation in the dry or imbibed state, and then subsequently maintained imbibed for different periods, have shown that post irradiation imbibition leads to fewer aberrations at germination (Villers, 1974). However, it is still an open question whether this signifies a DNA repair system operative during the period of imbibed dormancy or rather, a DNA repair that occurred after dormancy was broken, i.e., during the early hours of germination. If the latter is the case, then both in dry and in imbibed seeds the rate of loss of DNA integrity could be a major controlling factor in the rate of cell senescence in a quiescent seed. In the imbibed seed, however, the general maintenance of other cellular functions by continuous metabolic turnover would tend to sustain a fully effective cytoplasmic machinery and membrane compartmentalization not privileged to the seed stored dry (Chen and Varner, 1970).

Chapter 9

SENESCENCE PATTERNS OF FRUIT AND FRUIT RIPENING

Maturation refers to qualitative changes which allow the plant or organ to express its full reproductive potential. This is accomplished by a gradual transition of morphology, growth rate, and flowering capacity. The length of the maturation period can vary from days to years, depending on both genetics and the environment. One should be cautioned that maturation should not only be connected with flower formation but also with other reproductive structures such as tubers and bulbs. Although the period of maturation is genetically determined it can also be modified by plant growth substances/or environmental conditions. (Trippi, 1989)

The edible fleshy tissue of the commercially important fruit is derived from a diversity of tissues of the inflorescence. This group of fruit include both true fruit (such as the tomato) in which only ovular tissue develops into the fruit, and false fruits (such as strawberry) in which tissue of the receptacle of the flower develops into the bulk of the fruit.

Fruit maturation and ripening constitute a special type of senescence, in addition to the numerous types of leaf senescence classified of Simon.

In order to better understand the senescence process it is necessary to monitor quantitative changes which occur during the event. The ability to quantify specific regulating components which are central to the senescence process would be ideal; however, at the present time this is not possible due to the lack of information in this area. For now, it is necessary to use prominent components of the senescence process as measures. Components which have been used as measures of senescence include decreases in chlorophyll, total protein, nucleic acids, photosynthesis (reduction in photosynthetic enzymes such as ribulose bisphosphate carboxylase or phosphoenolpyruvate carboxylase), changes in plant growth substances, increased membrance permeability, and abscission. It must be noted that the study of senescence should be done with intact plants since there are many problems associated with the use of detached plant parts (Noodén and Leopold, 1978; Grover, 1993)

Fruits are similar to leaves in that they begin their development as green, photosynthetic tissues, at least in their exterior layers. The similarity is maintained to some extent during maturation and senescence-disappearance of chlorophyll and accumulation of yellow-red pigments is typical of both leaves and fruits. Nevertheless, the pigment changes associated with fruit maturation are generally more intense than those occurring during leaf senescence. The structural and biochemical differentiation of chromoplasts, for example, is usually more advanced and complete in mature fruits than in senescent leaves of the same species. The same can be said with regard to anthocyanins, which reach very high concentration in certain mature fruits.

In fruits that show a single sigmoidal pattern, ripening normally occurs during the final phase of slow growth of the fruit repening on the tree, there is probably no absolute separation in time between ripening and further growth processes and probably some small degree of cell expansion is occurring at least early in ripening. However, in those fruits that store large quantities of reserve materials, and which can ripen normally after detachment from the parent plant, there can be an almost complete separation of senescence from growth phenomena.

In addition to those fruits showing a single sigmodial growth curve, there are fruits showing more complex growth patterns with double and even triple sigmodial patterns. In these there are two or three rapid growth phases interrupted by one or two periods of little or no growth. The double sigmoidal pattern is shown by many fruits including the stone fruits, figs, and grapes (Table 9.1). It is thus clear that in those fruits showing double sigmoidal growth patterns active cell expansion accompanies some aspects of the ripening process. In the grape, the onset of ripening (veraison) is marked by the start of sugar accumulation loss of acidity, softening, and skin colouring;

these characteristic ripening changes are associated with renewed cell expansion.

Table 9.1: List of Fruits Showing Different Patterns of Growth

Single	Double	Triple
Apple	Apricot	Chinese Gooseberry
Avocado	Black currant	
Banana	Blueberry	
Date	Fig	
Mango	Grape	
Melon	Olive	
Orange	Peach	
Pear	Plum	
Pineapple	Raspberry	
Strawberry	Sour Cherry	
Tomato	Sweot Cherry	

In stone fruits, the two periods of rapid growth are associated with the development of different tissues. It is the tissues in which the ripening changes occur that are expanding rapidly in the period just prior to, and possibly during, ripening. It is clear that in the at least some fruits of this type it is not possible to separate senescence phenomena from those of growth, and this can lead to complications in the interpretation of the observed biochemical changes when both processes occur together.

FRUIT RIPENING PATTERNS

In view of the differences in the origin of the cells that ultimately constitute the edible parts of fruit, and the differences in the botanical origin and in the patterns of development of different fruits, it is perhaps not surprising that there is a range of patterns of behaviour during ripening. The processes of ripening generally occur in a relatively coordinated fashion in most fruits under natural conditions. An important factor to consider is that the modern varieties of fruit are often the result of continual selection by man, and they have features that make them very different from their ancestral wild types, their structure and composition has been largely determined by artificial selection rather than by the rigors of natural selection. The nature of the desirable changos in colour, flavour, and texture varies to some extend from fruit to fruit, and the desirable properties of the ripe fruit are characteristic of that particular fruit. The point during the sequence of ripening at which a fruit is ripe is, of course, to some extent a subjective evaluation.

Climacteric Fruit

The onset of the visible fruit ripening changes in apples was marked by an upsurge in the rate of respiration. The term "respiration climacteric" was coined to describe a critical phase in the life of the fruit. In detached apples, the respiration rate falls to a minimum value (the climacteric minimum) and begins to rise at the time of onset or ripening to a peak (the climateric peak) roughly twice the level of the minimum, and then declines slowly. The fruit reaches optimum eating quality some time after the climacteric peak.

The respiration in climateric fruits as the critical phase, separating maturation processes concerned with growth from the onset of the essentially irreversible changes of senescence. In the years following the pioneering efforts of Kidd and West, it became clear that the majority of both temperate and tropical fruits followed a similar climacteric pattern of increase in respiration during ripening. The extent and rapidity of the process varies with temperature and from fruit to fruit. In the banana at a normal ripening temperature the rise may be as much as tenfold, while in the apricot it may only be 30%. In many tropical and sub-tropical fruits, the climacteric peak is only a transitory one with a rapid decline following the peak and this is associated with a more rapid rate of ripening than in the apple. In the tomato the climateric peak is reached before there is appreciable accumulation of the red pigment, lycopene, and the rate of respiration is declining while the red pigment is accumulating. The climacteric rise in respiration was show by Kidd and West to occur even when the fruit remains attached to the tree, and this appears to be the case for nearly all of the fruits showing the climacteric pattern of ripening, and which ripen normally on the parent plant. An obvious exception is the avocado, for although it is a normal climacteric rise in respiration in detached fruit, attached fruits do not ripen and do not show a rise in respiration.

Non-climacteric Fruit

It became clear in the 1940s that not all fruits had the distinctive and well coordinated pattern of ripening of the typical climacteric fruit. The fruits may show neither a rise in respiration nor an accelerated production of ethylene during the ripening of mature fruit. The typical examples of this group are the citrus fruits, whose physiology has been studied in some detail. Citrus fruits are characterized by long growth periods and a very extended ripening process in which changes in chloroplast disorganization and pigment contents are often divorced from other ripening changes. Sound oranges produce small amounts of ethylene (0.02 to 0.06 al/kg/hr) throughout their development and maturation, but there is no massive rise in the production of ethylene associated with ripening as there is in the climateric fruits. The respiration of detached fruits declines slowly during storage, However, it was found that organs and lemons could be induced to produce enhanced amounts of ethylene and to give a rise in respiration if stored in an atmosphere of pure O_2.

STRUCTURAL AND BIOCHEMICAL CHANGES RESPONSIBLE FOR THE SYMPTOMS OF RIPENING

The visible signs of ripening are those changes in the color, texture, taste, and aroma of the fruit tissue that lead to the state in which the fruit is fit to eat, and these are determined by changes in the structural and metabolism of the fruit tissue. Ripening involves complex changes in metabolism of the fruit in which both anabolic and catabolic processes are involved. New pathways of metabolism are brought into play-which may lead, for instance, to the production of new pigments of new volatile compounds. In addition, degradative processes such as those involved in the conversion of chloroplasts to chromoplasts, and in changes in adhesion between cells leading to tissue softening also occur. These metabolic changes must reflect changes in the in vivo activity of the enzymes involved and the change in activity may involve the synthesis, degradation, release, activation, or inactivation of these enzymes. These changes, at least in the climacteric fruit, have to be considered in relationship to the changes in respiration and in ethylene production. The underlying structural and biochemical changes responsible for the visible changes of ripening will now be considered.

Changes in Cell Structure

The edible tissue of fleshy fruit is composed usually of parenchymatous cells, frequently of very large size. The increase in ploidy may be a factor determining the large size of fruit cells. These cells have a high water content and accumulate organic substances such as sugars and organic acids. In addition to the large parenchymatous cells, the flesh of many fruits contains stone cells, which may play a role in the texture of the ripe fruit and which contribute to the taste of some fruits. Since they frequently contain tanin-like compounds, leucoanthocyanins, which are responsible for the astringency of these fruits. The major ultrastructrual changes occurring during ripening occur in the plastids. Oeller et al(1991) showed that mature green tomatoes contain chloroplasts with relatively small grana and frequently with one or two vacuoles bearing starch. The first signs of the onset of ripening changes are the appearance of electron dense regions in the granal and intergranal membranes which are though to represent the initial stages in lycopene deposition, and an increase in the number of plastoglobules. Later the inter-granal region swells and the granal diameter increases markedly. By the time the orange stage is reached, new structural called thylakoid flexes appear; these resemble the loose type of prolamellar body found in plastids in etiolated leaves exposed to light for short periods. Lycopene crystalloids of up to 18 a in length appear by the orange-red stage, and crystalloids composed of concentrically arranged lamellae are frequently found. In the later stages of ripening there is a reduction in the amount of membranous material in the stroma and a large increase in the size and number of the plastoglobules.

Other cell components show less dramatic changes during ripening. All the nonplastid cytoplasmic structures observed in the cytoplasm of immature tomatoes are also observed in the ripe fruit. These structures include mitochondria, Golgi bodies, ribosomes, endoplasmic reticulum, small vacuoles and vesicles, microbodies, and lipid droplets.

Factors Responsible for the Changes in Colour During Ripening

A major factor in the colour changes of fruit during ripening is the transition from chloroplasts rich in the green pigment, chlorophyll, into chromoplasts rich in red or yellow carotenoid pigments.

Effect of photosynthetic CO_2 fixation

The metabolic significance of the transition is the loss of the potential ability to carry out photosynthetic CO_2 fixation. It has been shown that in apples the chloroplasts of the peel are functionally active in photosynthesis, and even the internal tissues of the pulp contain active plastids. Experiments using $^{14}CO_2$ showed that in the light CO_2 was fixed into normal products of photosynthetic activity was constant during the early part of the climacteric rise in respiration, but decreased during the later stages. The ratio of respiration to photosynthesis, at the late stages in ripening. There was no complete parallelism between the loss of chlorophyll and the loss of photosynthetic activity. The growth continued in the dark and at harvest was only 6% less than that of control fruit. Development in darkness had no effect on the total malate content of the fruit, but there was less malate in the outer layers of cells compared with control fruit. It thus seems clear that the limited photosynthetic activity of fruit has only a small effect in promoting the accumulation of reserve materials such as malate; this is an line with the view that most reserve materials are transported to the fruit other parts of the tree, principally from the leaves. Thus $NADPH_2$ and ATP, do not apear to play an important role in the regulation of the climacteric and ripening.

Changes in pigments

There are also changes in the pigments that make an important contribution to the colour of fruit.

Chlorophylls

Investigators of pigment changes in senescing fruits and leaves diverted most of their efforts towards carotenoids and anthocyanins, leaving the in vivo degradation of chlorophylls least understood. Most investigators were unable to find any coloured degradation products, except for diminishing amount of chlorophyll a and b. The seemingly inevitable conclusion was that chlorophyll undergoes rapid, probable enzymatic cleavage into small, colourless fragments. Removal of the phytol by chlorphyllase has been suggested to the primary step in chlorophyll degradation. A marked rise in chlorophyllase activity has, nevertheless, been observed recently during ethylene-induced degreening of citrus fruits. The precise role of chlorophyllase in vivo awaits further clarification.

Chlorophyll a/b rations have been found, with few exceptions to decrease slightly during leaf senescence suggesting a somewhat faster degradation of chlorophyll a. This pattern may become meaningful if one considers the distribution of chlorophylls a and b in various chlorophyll-protein complexes. The fate of these complexes during senescence has not yet been determined.

In vivo absorption spectra taken did not reveal qualitative difference between green and senescent *Rumex* leaves. The presence of pigmented degradation products of chlorophyll has, nevertheless. It has been indicated recently by spectrofluorimetric techniques.

In summary, the in vivo destruction of chlorophyll is still an open, intriguing problem awaiting elucidation.

Carotenoids

Carotenoids are normal constituents of the photosynthetic apparatus an are, therefore, present in green fruit tissues prior to maturation. Maturation does not always involve an accumulation of carotenoids- many varieties of apples, pears, grapes, and olives belong to this group. The yellowing of mature bananas also results from preexisting carotenoid unmasked due to the disappearance of chlorophyll.

On the other hand, numerous fruits accumulate large amounts of carotenoids during maturation. Young fruits contain mainly foliar carotenoids like lutene and neoxanthin, which decline towards maturation simultaneously with chlorophyll.

Carotenoids are usually biosynthesized during the later stages of maturation,, but the complement of pigments differs greatly from one species to the other. Mature citrus fruits, on the other hand, contain about 115 different carotenoids, which is about on third of the total carotenoids known to occur in nature along with the increase of total xantohphylls in orange peel.

The pathway of either breakdown or the interconversion of carotenoids cannot be, at present developed into a single scheme which fits all plant tissue.

Anthocyanins

The normal forms of anthocyanins in fruits are ß-glycosides of six principal aglycones (anthocyanidins). esterification of the sugar hydroxyls with organic acids yields acylated anthocyanins, which appear in significant amounts in grapes. Other phenols, such as tannins and additional flavanoids, may interact with anthocyanins in a variety of ways to yield effects of copigmentation of discolouration. Phenols and their condensation products appear also as pigments in their own right in dates.

The external expression of anthocyanin pigmentation also involves pH effects and complexes with metals. At low pH levels anthocyanin solutions are red, at pH levels near and above neutrality they become blue or purple. The vacuoles in which may anthocyanin are contained are believed to be at low pH levels. The blue appearance of fruits must a attributed to metal complexes and copigmentation.

The biosynthesis of anthocyanins during maturation appear to intimately related to carbohydrate accumulation. The link between carbohydrates and anthocyanins may be via the pentose phosphate pathway, which leads through shikimated to anthocyanin.

FACTORS RESPONSIBLE FOR THE SOFTENING OF FRUIT DURING RIPENING

Changes in the texture of fruit during ripening result from changes in the structure and composition of their cell walls. The cells of fruit are frequently large parenchymatous cells with large air spaces between them. Changes in the degree of contact between cells, as well as in the structure of the cell walls themselves, will affect texture. During ripening in the apple there was a decreases in the galactose content of cell wall hydrolysates. In addition there is an increase in water-soluble polyuronide, which has been correlated with cell separation and the softening of the tissue. An increased solubility of the cell wall polyuronides is the characteristic cell wall change occurring during softening in a whole range of ripening fruits.

Softening by Enzymes

A number of enzymes attacking polyuronides have been detected in ripening fruit. Frequently the activity of such enzymes is low or absent in unripe fruit and increases during the ripening process.

Effect of Polygalacturonase PG

The enzyme frequently invoked as being involved in the solubilization of pectin is the cell wall-bound polygalacturonase (PG). Which acts on de-esterified pectin molecules and attacks linkages between galacturonic acid groups in the polyuronide. Two types of PG have been found in fruit; one, an exo-PG, acts by cleaving free galacturonic acid groups from the non-reducing end of the molecule and the other, an endo-PG, which acts by cleaving the chains of polyuronides in a random fashion. It is suggested that the solubilization reaction depends on the rupture of linkages between the polyuronides and other structural components of the cell wall, rather than extensive degradation of the polyuronide chains.

Effect of pectin methyl esterase

If PGs play a role in the increase in the water solubility of pectins during ripening, then it is likely that a second enzyme, pectin methyl estease (PE) also play a role. The polyuronides of the middle lamella are highly methylated, while PG is relatively inactive towards methylated polyuronides and requires free carboxyl groups for its action. PE is a wall-bound enzyme that can be readily desorbed from cell walls in vitro by addition of neutral salts at slightly alkaline pH, and its role may be to modify pectin structure prior to PG action. There is some evidence for a decrease in pectin methylation during ripening and it may well be that quite limited PE activity could greatly promote PG activity and softening. PE can exist in multiple forms in ripening fruit. In general, the potential PE activity far exceeds the net change in methylation observed during ripening, so that even if it is indeed involved in the softening reaction the factors controlling its activity in vivo are far from understood.

The role of cellulases which cataylase the breakdown of cellulase in fruit ripening is controversial. In avocado, cellulase activity rises during the climcteric to very high values and cellulases may be synthesized or activated very early in the climacteric and play an important role in the early stages of softening. In other fruits, however, cellulases appear to play only a minor role in softening. It is clear that changes occurring in the softening of fruit tissues during ripening are extremely complex, so it is perhaps not surprising that there is still doubt about the role the various enzymes play in the softening reaction.

INTERRELATIONSHIPS OF THE DIFFERENT ASPECTS OF THE RIPENING PROCESS

Two views of interrelationships of different aspects of ripening process exist. One considers the rise in respiration as a response to the increased energy demands made on the fruit by the various ripening changes. There is now, however, considerable evidence that the many aspects of ripening are not dependent on the respiration rise. For instance, the application of inhibitors of protein synthesis such as cycloheximide to unripe pears and bananas inhibited the production of ethylene, softening, and loss of chlorophyll during subsequent storage without inhibiting the rise in respiration. In bananas, in which ripening had been initiated by a short exposure to ethylene, subsequent storage in a low oxygen (0.5 to 1%) atmosphere containing ethylene accelerated the change in fruit colour in the absence of a rise in respiration. These results show that it is possible to isolate many aspects of ripening from the rise in respiration, and they clearly throw doubt on the view of ripening as a highly integrated process. The view has been expressed that the respiration climacteric is neither dependent upon nor integrated with other aspects of ripening.

An alternative view of the role of the respiration climacteric has been proposed according to which the activity of the mitochondrial system as operating relatively independently of the other ripening changes. The rise in respiration should be regarded, in Romani's view, as just another independent ripening change and this be related to the quasi-autonomous status of mitochondria in cells. Studies on the properties of mitochondria isolated from fruit at various stages of ripening suggest that, although there may be changes in the mitochondria composition such as changes in the ability to metabolize oxaloacetic acid during the climacteric, mitochondria remain fully functional throughout the period of the respiratory rise and beyond. Isolated mitochondria oxidize substrates at increasing rates during ripening, and retain the ability to phosphorylate rates during ripening, and retain the ability to phosphorylate ADP with a high respiratory control ration throughout ripening. Indeed the evidaence appears to suggest that there is an increased capacity for phosphorylation as measured by ADP/O ratios as the fruits ripen. Romani suggested that the increase in respiration is a response to changes occurring in the cytoplasm, the nature of which is unknown, but increased availability of cofactors and substrates or removal of some of the other constraints limiting in vivo activity of mitochondria could be involved.

It is clear from the preceding discussion that at least some of the ripening changes can be experimentally dissociated from each other. However, in the field, under normal conditions, ripening does occur as a well-coordinated sequence of changes.

FACTORS RESPONSIBLE FOR THE CHANGES IN FLAVOUR OF FRUIT

It general, fruits show a decrease in acidity and an increase in sweetness during ripening. There are, however, some exception to this general rule such as the banana, in which the pH falls during ripening from 5.4 to 4.5. and this is associated with an increase in the content of organic acids such as malate and citrate. The absolute levels of sugar and acids, as well as the ratio of sugar to acid, play an important role in taste of the ripe fruit.

The pH of the sap of fruit cells is frequently below pH 7 and can be as low a pH 3.0 in lemon cells. Two acids are most frequently found in fruit cells; these are malate and citrate. While these are the major acids, many fruits store a whole range of other acids of oxidation of the malate, it is clear that factors altering the permeability characteristics of the tonoplast membrane can have an important role to play in utilization of organic acids.

In addition to the changes in sugarid balance, an important factor in the flavor of fruits is their aroma, which arises from the production of volatile compounds by the fruit during ripening. Even compounds of low volatility can contribute to the aroma of fruits on account of the extraordinary sensitivity of the olfactory epithelium tissue of the nose to some compounds. These volatile compounds include many different classes of organic molecules such as organic acids, alcohols, esters, carbonyl compounds, hydrocarbons, terpenoids, and there are wide differences in the contributions that individual compounds make to the resulting aroma, which are not simply related to their rates of production.

The production of most volatile odour compounds is generally very low in unripe pre-climacteric fruit, but it rises many-fold during ripening.

The formation of the volatile compounds, trans-2-nonenal, trans-2-cis-6-nonadienal and trnas-2-hexenal, which are important constituents of cucumber flavour have been studied. These important flavour constituents are not present in vivo, but are formed during chewing. Much remains to be done on the enzymology of the endogenous volatile production of ripening fruits: the description of the properties of the individual enzymes and the changes in their activity during ripening will be a necessary prelude to a study of their regulation during ripening.

FRUIT RIPENING: BIOCHEMICAL CONTROL

While there is uncertainty about the mechanism by which ripening is initiated, there is equal uncertainty about the nature of the underlying biochemical events that result from this initiation and that lead to the visible symptoms of ripening. In the climacteric fruits, ripening is seen as the final phase in development of the organ and, as such is under genetic control. There is a sharp transition in metabolism at the time of the onset of the climacteric and this involves the promotion of both anabolic and catabolic processes. Two theories have been advanced to explain the underlying biochemical control of the levels of enzyme activity responsible for the promotion of these processes. In one, the important feature is the de novo synthesis of a small group of specific enzymes, which catalyze critical steps in the changes leading to ripening, but which are not present before the climacteric. Inherent in this theory is the synthesis of new messenger RNA species for these enzyme protein. The alternative theory suggest that the activation or release of pre-existing enzymes is the dominant feature and this is linked with changes in the compartmentation of the cell, which could lead to the release of hydrolytic enzymes from lipoprotein-bound cell organelles and to changes in the permeability of cell membranes, leading to increased availability of substrates, cofactors or activators to the active site of the enzymes. No conclusive evidence for either theory has yet been advanced, but it seems likely that both types of change can operate in the various ripening processes.

Evidence for the first theory was presented by the work of Hulme. He showed that in detached apples the protein nitrogen content rose during ripening. However, for a number of climacteric fruits such as applies, avocados, pears, and tomatoes there is evidence for a period of enhanced incorporation of amino acids into proteins during the early stages of the climacteric. As in the case of protein, so with RNA there is evidence for a period of enhanced incorporation of RNA precursors such as uridine into RNA at the early stages of the climacteric in apples, avocados, pears, and tomatoes. There is good evidence for a period early in ripening in which both RNA and protein synthesis are active.

Further evidence for the importance of protein synthesis in ripening has come from the use of inhibitors of protein synthesis. There are many reports of increases in the in vitro activity of enzymes extracted from fruit at various stages of ripening, but often it is unclear whether these result from changes in the fruit changes as it ripens, from changes in the from and activity of the isolated enzymes or from changes in the tissue concentration of the enzyme. Although an increased rate of amino acid incorporation is a feature of the early stages in ripening in a number of climacteric fruits, there are few reports of studies on the nature of the protein products of the incorporation. The incorporation of amino acids into proteins during the early stages of ripening represented the turnover of a substantial proportion of the proteins of the pulp cells, and that the incorporation was into a wide range of soluble proteins rather than into a few noval proteins specifically involved in ripening. On the other hand, it was found that there was a peak in the incorporation of tritiated uridine into RNA just before the onset of colour changes in ripening tomatoes. It was concluded that at about the time of onset of ripening tomatoes. It was concluded that at about the time of onset of ripening there was a change in the composition of the m-RNA fraction which could be involved in directing the synthesis of a few novel enzymes involved in ripening.

These conflicting reports leaves the significance of the role of RNA and protein synthesis in ripening unclear. Thus it was concluded that there was a change in the permeability properties of the cell membranes during ripening. Such changes in permeability could clearly have a profound influence on the activity of enzymes by making substrates, cofactors, activators or inhibitors more freely available to them.

Changes in membrane permeability have been estimated by measurements of changes in apparent free space (AFS) to organic molecules and to inorganic ions, and by measurments of changes in the leakage of solutes from tissue slices.

The measurements of changes in AFS and of solute leakage are relatively crude measures of changes in the permeability characteristics of the cell. They give no information on changes in compartmentation within the cytoplasm, but give some measurement of tonoplast and/or plasmalemma membrane permeability.

It is of interest to know such permeability changes, whether they involve active or passive transport, might affect the activity of enzymes involved in ripening. An example of this has been proposed in a study of the activity of the enzyme phospofructokinase (PFK). The activity of phosphofructokinase measured in vitro does increase during the ripening of either bananas or tomatoes, yet there is good evidence from the accumulation of its product, fructose 1:6 bisphosphate, during ripening of bananas and tomatoes that glycolysis is activated during the climacteric rise in respiration, and that the in vivo activity of PFK increases markedly during this period.

This discussion shows that ripening is a differentiation process that is ultimately determined at the genome level. The process by which this control is expressed ultimately as changed activities of enzymes and pathways is unknown. The extent to which the increased enzyme activities responsible for these changes are due either to de novo synthesis, activation, or release of enzymes from membrane-bound structures cannot be assessed, although all three processes may be involved. It is clear that increased turnover of proteins occurs, at least in bananas, as a direct response to ethylene and this may obscure the formation of new enzymic proteins at this stage. In some cases, at least, membrane permeability changes appear to occur during ripening, and we know very little of the particular membranes that are affected and can only speculate on the possible metabolic significance of such changes.

RIPENING: HORMONAL CONTROL

The use of Plant Growth substances to control fruit set, size, shape and maturation has become important in agriculture today because they have the ability to increase fruit size, color, and shape, thereby increasing marketability. In addition, by hastening or delaying maturation the grower can utilize peak demands, avoid unfavourable environmental conditions and extend the market period.

Some plants, e.g., the cultivated banana, naval orange, oriental of fig and pear, develop fruits in nature in the absence of pollen, other plants, e.g., certain grape varieties, develop fruit from the stimulus provided by the presence of pollen alone without fertilization of the ovule, presumably because of its secretion of hormonal substances.

Synthetic growth regulators that cause development of fruits in plants include 4-chlorophenoxyacetic acid and 2-naphthoxyacetic acid. These chemicals are most effective on fruits with many ovules, e.g., tomato, squash, eggplant, and fig. However, these chemicals usually are ineffective on peach, cherry, plum, and other stone fruit. Many fruits that can be set by such hormonal compounds also can be set by gibberellins. Gibberellins also can set fruit in some species that do not respond to the other growth regulators. An extensive study has shown gibberellic acid (GA3) to be the growth regulator of choice to improve fruit set of citrus. Application of aqueous sprays of gibberellic acid to entire trees in full bloom increases the set and the yield of seedless fruit of five self-incompatible citrus cultivars. Seedless fruits, however, are more likely to drop after initial fruit set than seedy fruits produced by cross-pollination. Gibberellins have been used in the production of seedless Dclaware grapes in Japan. One of the single largest uses of GA3 in the United States is in the production of Thompson seedless grapes.

Yield improvement from cucumber plants is directly related to increased fruit set on individual plants. Generally, cucumbers do not set more than two fruits per plant at any one time because the earliest fruit that sets restricts the development of additional fruits. Certain growth regulators, e.g., methyl chlorflurenol, effectively overcome this limitation in fruit set. However, for chlorflurenol to be totally effective, the plant must have predominantly female flowers. Ethephon induces female flowering in the cucumber and enhances the potential for chlorflurenol to increase yield.

The ability to control the shape of fruits has practical implications. For example, apple cultivars known to have superior internal quality attract premium prices provided the consumer can identify the product by one characteristic feature of its external appearance. In the market place, the Red Delicious apple is identified primarily by the elongated appearance of the fruit and the presence of pronounced calyx lobes on the crown. A fruit having these attributes is called *typey*. Commercial application of a growth regulator product that is a combination of gibberellins A4 and A7 with the cytokinin (6-benzyladenenine) causes improved typiness Red Delicious apples. The period between full bloom and early petal fall is when the flower parts are most receptive to treatment and capable of absorbing and transferring the regulators to the developing fruit. The success with apples has resulted in this product being tested for a number of different uses on a wide variety of crops.

Auxins and Pollen

Both pollination alone and growth of pollen tubes in the style can stimulate the growth of the ovary, even if fertilization does not occur.

Fitting (1909) showed that application of aqueous pollen extracts to orchid flowers stimulates overy growth, indicating that the stimulation is caused by a chemical substance that fits the definition of a hormone. Other investigator (Yasuda, 1984; Gustafson, 1937) later showed that pollen extracts of many unrelated genera can stimulate ovaries to develop fruit. Gustafson (1936) proved that auxins are responsible for the stimulus of pollination by demonstrating that synthetic auxins can induce unpollinated ovaries to develop into fullsized fruits. One theory is that an activator contributed by the pollen results in an enzymatic release of auxin in the tissues of the ovary (Muir, 1947). It may also be that the process of fertilization itself is the main stimulus that results in the high concentrations of auxin found in the fertilized ovary.

Auxins and Fruit Set

Many synthetic auxins set fruit in plants. Best set is usually obtained with 4-CPA or BNOP. Auxins are most effective on fruits with many ovules, such as fig. strawberry, squash, tomato, tobacco, rose, and eggplant. Auxins are ineffective on peach, plum, cherry, and other stone fruits. In all only 20 per cent of the horticultural crops have responded to auxins, indicating the hormones other than auxins must affect fruit set.

Auxins and Fruit Growth

Increase in fruit size is caused mainly by cell enlargement. It is then not surprising that since auxins control cell extension, they have been though to play a dominant part in determining the growth patterns of fruits. Two main lines of evidence support this hypothesis. First, there is a correlation between seed development and final size and shape of fruit; second, application of auxin to certain fruits at particular stage of their development induces a growth response.

Seed Development in Relation to Size and Shape of Fruit

Fruit size and shape are closely corrleated with seed number and seed distribution in many fruits. The endosperm and embryo in the seed produce auxin, which moves outward and stimulates growth of the endosperm development (Nitsch, 1950). The location of seeds also greatly influences or controls fruit shape.

Changes in Patterns of Fruit Growth Caused by Auxin Application

The hypothesis that auxins control fruit growth is strengthened by the fact that applications of synthetic auxins can enlarge many fruits or change their pattern of growth. The effect of auxin application on increase in fruit size has been used to commercial advantage with several plants, such as blackberry, grape, strawberry, and orange. A striking example of the change that can be effected in the fruit growth pattern of fig was demonstrated by Crane (1949). When 2,4,5-T was applied to the fruit of fig (which has a double sigmoid curve) at the beginning of the second stage of growth, the fruit continued to grow and ripened in 60 days instead of the normal 120 days.

Influence of Seeds on Fruit Growth

Developing seeds usually have an influence on fruit growth. Convincing evidence for this fact has been presented by Nitsch (1950), who experimented with cultivated strawberry (Fragaria spp.). When achenes were removed from the area of a receptacle, growth ceased in that area. Growth occurred around areas where achenes were retained. If BNOP or another auxin is applied to replace the effect of the achenes, full-sized fruits are obtained.

In other fruits seeds are required for growth during only part of the growth period. For example, fruit growth in apple is dependent on the presence of growing seeds up to the end of the June drop (Abbott, 1959). After this stage, removal of the seeds no longer results in a cessation of fruit growth. In later stages of fruit development, tissues other than those in the seed probably synthesize auxin. In the seedless 'Washington Navel' orange and 'Black Corinth' grape, fruit development has become completely independent of seeds.

Production of Auxin in Seeds

Ovules and young seeds are the main sources of auxin. Gustafson (1939) demonstrated that the concentration of auxin in the seeds of immature fruits of tomato was higher than that in the placenta and that the concentration of auxin in the central axis and in the partition of the fruit was in turn higher than that in the carpel wall. These data would seem to indicate that auxin is produced in the seeds and then diffuses outward along a concentration gradient to other parts of the fruit. In apple the embryo also contains a considerable amount of auxin although not as much as the endosperm (Luckwill, 1948).

Correlation between Auxin Content of Seeds and Fruit Growth

Although close correlations frequently exist between number of seeds and final fruit size and between seed distribution the shape of fruit, there is usually no close correlation between the total amount of auxin produced in the seeds and fruit growth.

It could be hypothesized that growth is not related to the total auxin complex but may be controlled by one or more specific auxins in the complex. Wright (1956) found that two of four auxins correlated well with fruit growth in the European black currant. To date, however, there has been little evidence in support of this idea.

Stahly and Thompson (1959) found three auxins in peach seeds during growth. Auxin I was present in very small quantities and showed no correlation with growth. However, auxins II and III, which were produced by the endosperm and embryo, respectively, were present in largest amounts during the period of slow growth of the fruit. Stahly and Thompson (1959) suggest that the slow growth period was caused by auxin concentrations that were high enough to inhibit cell enlargement. Although these investigators believe that the decline in growth and the accompanying peak in auxin concentration were correlated with a peak in embryo growth. The general lack of correlation between endogenous auxins and fruit growth suggests that, for some species, auxins do not control fruit growth. The conclusion follows that other classes of plant regulators also play a part.

GIBBERELLINS AND FRUITSET

Many fruits that can be set which auxins can also be set with gibberellins. However, gibberellins have also been effective for setting fruit in several species that do not respond to auxins. Gibberellins produce a good set in tomato but the fruit remains small.

Attempts to induce parthenocarpy in apple by applying auxins have produced rather poor results, but gibberellins have yielded positive results (Luckwill, 1959; Dennis and Edgerton, 1966). However, the induced fruit set has been generally low, and application of gibberellins to open-pollinated trees at petal fall has resulted in reduced yields.

There is a considerable specificity among the gibberellins in promoting parthenocarpy. Bukovac and Nakagawa (1967) treated emasculated flowers of 'Wealthy' apple with gibberellins GA1 through GA10 and GA14 by applying labolin pasts @ 10^{-3M} to the cut style and adjacent receptacle tissue. Nonpollinated controls did not enlarge and abscissed within two or three weeks after treatment. After four weeks GA4 and GA7 showed the best activity; the growth rate of fruit treated with them was equal to that of pollinated controls. GA3, GA6 and GA8 showed tha least activity and GA1, GA3, GA4, GA10, GA13, and GA14 evidenced intermediate activity. At maturity all persisting parthenocarpic fruits were approximately the same size as seeded fruits, except for those in which parthenocarphy had been induced by GA5 and GA14. It is not surprising that GA4 and GA7 are very active since they have been found to be present in immature seeds of apple (Dennis and Nitsch, 1966).

A mixture of GA4 + 7 at a concentration of 400 ppm also increased fruit set of the emasculated 'Sturmer' variety of apple is New Zealand and the Red Delicious variety in Washington State, but not 'Golden Delicious' variety (Williams and Letham, 1969). The GA4 + 7 acted synergistically with the cytokinin PBA at a concentration of 400 ppm in increasing set in 'Sturmer.' Cytokinins alone enhance parthenocarpic fruit set in some varieties of apple but are less effective than gibberellins.

Dennis (1967) concluded that the gibberellins that are produced in the ovule after fertilization are responsible for fruit set in apple. However, the role of gibberellins in the early stages of fruit development is not yet clear because Luckwill and his colleagues (1969) failed to find gibberellin activity until five weeks after pollination.

GA3 usually has little effect on fruit set in seeded commercial varieties of apple unless pollination is prevented. Dennis (1970) concluded that seedlessness alone is not correlated with responsiveness to gibberellin application.

Auxins are ineffective in setting citrus, but gibberellins have shown promising results. Coggins et al. (1966) demonstrated the possibility of increasing fruit set by applying gibberellin to lemon, lime, and other varieties of citrus, but unfortunately application of sprays to whole tress caused phytotxic effects. All mandarin fruits dropped before maturity. These investigators suggest that, in spite of the labor involved in emasculation, it should be possible to produce parthenocarpic grapefruits, at least on a restricted scale, in home gardens.

The two growth regulators together resulted in sets of 94 percent set. GA3 alone and NAAm alone resulted in sets of 70 and 65 per cent, respectively.

Gibberellin has effectively stimulated parthenocarpy in several stone fruts, including apriot, peach, and almond (Crane *et al.,* 1960, 1961), fruits in which negative or, at best, poor results were obtained with auxin. The size of mature parthenocarpic fruits was found to be approximately the same as that of open-pollinated fruits. Parthenocarpy has also been induced in the 'Sultan' plum by application of GA3 and mixture of GA and GA (Jackson, 1968); the mixturo was found to be more effective. Although parthenocarpic fruits of plum grew more rapidly early in the season, their final diameters were only about 60 percent of the controls.

Gibberellin induces parthenocarphy in emeseulated clusters of both seedless and seeded varieties of *Vitis vinifera* grape, although the final berry size is somewhat smaller than that of open-pollinated clusters (Weaver and Sachs, 1968). The set of open-pollinated clusters is usually decreased.

Mango is a very difficult plant in which to induce fruit set. Chacko and Singh (1969), working in India, successfully induced fruit set in mango cv. 'Dashehari' by applying BA followed by of BNOA at a concentration of 10ppm and GA3 at a concentration of 250 ppm.

Gibberellins and Fruits Growth

Young seeds are rich sources of gibberellin-like substances. Such substances have been isolated from the seeds of plum, apricot, almond, and young Phaseolus, for example. More gibberellin is usually present when seed growth is rapid. Bound gibberellins have also been found in some seeds, and often several gibberellins are present. MacMillan et al. (1961) identifed GA_1, GA_3, GA_6, and GA_8 in immature seeds of runner bean.

Correlation Between Endogenous Gibberellin and Fruit Growth

As with auxin, little correlation has been found between the gibberellin content of fruit and fruit growth. The concentration of gibberellin-like substances in the seed, endocarp, and mesocarp of apricot correlates closely with the growth rates of these tissues between anthesis and maturity but not with general fruit growth (Jackson and Coombe, 1966); Iwahori et al. (1968) studied the gibberellin activity in berries of two cultivars of *Vitis oinifera* grape- "Takay'. a seeded variety, and 'Seedless Tokay'. The seedless variety developed as a mutant of the seeded plant. These investigators found that gibberellin-like activity begins shortly after fertilization, when berries are young and rapidly growing. In Seedless Tokay seed development ceases at an early stage. The earlier decrease of gibberellin content in 'Seedless Tokay' is probably caused by the abortion of the embryo and cessation of seed development. A close correlation was found between the gibberellin-like activity and berry growth during stages one and two but not in stage three.

Research done by Jackson (1968b) in New Zealand prompted him to conclude that the gibberellin concentration in peach tissues is closely correlated with the rate of cell expansion in each tissue but not with cell division or general fruit growth.

Changes in Growth Pattern Induced by Application of Gibberellin

Application of gibberellin, but not auxin, induced parthenocarphy in *Rosa arvensis* (Prosser and jackson, 1959). Two acidic auxins and one neutral auxin were found in fertilized or normal fruits (hips). The growth pattern of the parthenocarpic fruits produced following application of GA3 was similar to that of normal hips, even though the former contained none of the three. Similar results were obtained with the 'J.H. Hale' peach by Crane et al. (1960). The growth pattern of parthenocarpic (seedless) fruits produced by gibberellin closely paralleled that of the cross-pollinated fruits.

Gibberellin application anlarge the fruits of several species, including most seedless varieties of grape. The time of application also has a marked effect on berry shape (Christodoulou et al. 1968; Zuluag, 1968). To date most work has been done on the Thompson Seedless' and 'Black Corinth' varieties. Gibberellin treatments applied at the time of bloom elongate the berries and that later applications produce larger barries (Christodoulou *et al.,* 1968). Application of gibberellin resulted in the greatest increases in berry size of seedless varieties; the amount of increase declined as seed number increased from one to four.

Gibberellin and IAA have a synergistic effect on the growth of tomato fruits. Fruits more than twice as large as those obtained by application of either hormone alone were obtained (Luckwill, 1959a), indicating that both hormones affect fruit growth.

There is also evidence that the effect of gibberellin on fruit growth may be to increase the amount of auxin in the ovaries.

Asymmetric fruit growth can be induced in seeded and parthenocarpic fruits of apple and Japanese pear following a localized application of GA4 and GA7 to one side of the fruits two weeks after bloom (Nakagawa *et al.,* 1968). The number and size of cells of the treated sides increased: cells or the nontreated sides remained unchanged.

Seedless fruits of apple that are induced to set by gibberellins usually produce much more elongated fruit that do normal seeded fruits. Similarly, fully seeded fruits of apple respond to application of GA3 by producing longer fruit. Just as do the parthenocarpic fruits set by the hormone (Westwood and Bjornstad, 1968). Application of GA3 has also bean reported to increases the depth of the stem cavity of 'McIntosh' apple (Webster and Crowe, 1969).

Application of gibberellin develops parthenocarpic fruits of peach and fig that are similar to those resulting from pollination (Crane, 1964; Stembridge and Gambrell, 1970). Application of gibberellin has also resulted in production of mature parthenocarpic pears (Griggs and Iwakiri, 1961).

CYTOKININS AND FRUITSET

Application of BA and PBA was found to be effective in increasing set in open-pollinated clusters of two seedless varieties ('Black Corinth' and 'Thompson Seedless') and three seeded varieties ('Muscat of Alexandria, ' "Tokay', and 'Almeria') of *Vitis vinifera* (Weaver *et al.*, 1966). However, the more soluble and mobile PBA was always more effective than BA. Both auxin and gibberellin induced parthenocarpic set in the Calimyrna ('Lob Ingir'). Fig and Crane (1965) demonstrated that PBA is also effective. Treatment with BA is also useful for increasing fruit set in muskmelon (Jones, 1965). It is well known that cytokinins increase DNA, BNA, and protein synthesis and that they can mobilize metabolites to the area of application of a compound. Jones believe that the action of BA in muskmelon is to increase the competitive ability of the treated fruits. Cytokinins are also effective for setting fruit in emasculated flowers of some varieties of apple, although they are generally less effective than gibberellins (Williams and Letham, 1969).

Cytokinins and Fruit Growth

Developing fruits are rich sources of cytokinins, which are found in tissues where rapid cell divisions are occurring. The highest concentrations have been found in young fruits, particularly the seeds. Active extracts have been obtained from fruitlets of apple, quince, pear, plu, peach, and tomato (Letham, 1967). Cytokinins play an important role in the regulation of cell division in fruits of apple and plum and probably in most other furits as well (Kaminek *et al.*, 1992).

Effect of Synthetic Cytokinins

Much of the research in this area has been done with BA and PBA. Crane (1965) treated 'Calimyrna' fig with PBA at a concentration of 500 ppm and obtained parthenocarpic figs similar to those obtained by application of an auxin or a gibberellin.

The berry size of the seedless 'Black Corinth' grape was increased by application of PBA at a concentration of 1,000 ppm (Weaver and van Overbeek, 1963), but the berry size of the 'Thompson Seedless' variety was only slightly increased (Weaver *et al.*, 1966). When applied to such seeded varieties as 'Almeria,' PBA prodces many small berries that fail to develop BA is usually less effective than the more soluble and mobile PBA. The results of the experiments performed with grape would seem to indicate that cytokinins exert most of their effect on fruit set, gibberellins affect mostly fruit growth, and auxins affect both set and growth.

Cytokinin treatment causes fruits of apple to enlarge and to have well-developed calyx lobes. When 'Delicious' apples were treated with N-(purin-6-yl)-x-phenylaglycine. PBA, BA, and zeatin at concentrations of 100 to 500 ppm four days after full bloom, all compounds stimulated development of elongated fruits with prominent, well-developed calyx lobes. This gave the fruit a knobby apperance and a more open calyx (Williams and STahly, 1969). Letham concluded that shape of the apple at maturity probably depends on the balance between gibberellins and cytokinins in the fruitlets and that apple variety differ in there response to these compounds.

ETHYLENE AND FRUITSET

Ethylene usually causes abscission of flowers and young fruits, so its effect on fruit set is frequently a negative one.

Ethylene and Fruit Growth

Ethylene is now considered to be one of the main plant hormones. Many plant responses that were formerly believed to result from the presence of auxins are now ascribed to induce ethylene production (Burg and Burg. 1966).

Typical flavour did not develop in fruit that were treated in the first half of stage two, but normal flavour developed in those treated during the last half of stage tow or during stage three. Maxie and Crane hypothesized that ethylene initiates growth during stage three and stimulates maturation of fig fruits. Crane et al. (1970) also demonstrated that application of ethephon hastens maturation of the fig.

The research on fig prompted the testing of ethephon on peach. Fruits treated with compound at a concentration of 100 ppm measured 2.91 inches in diameter compared with 2.49 inches for the control fruits. Higher concentrations of the compound caused severe abscission of both fruit and leaves. Concentrations of 12.5 to 100.0 ppm also increased the carotenoid content of harvested fruits.

Ethylene as a Ripening Hormone

There were two schools of thought regarding the role of ethylene in fruit maturation :

Ethylene as a ripening hormone (Kidd and West, 1945).

Ethylene is a by-product of fruit ripening (Baile 1960) and it is a by-product of fruit maturation. According to by-product theory, the internal ethylene content with the rate of ethylene production, one can advance the argument that small quantities sufficient to induce ripening are produced prior to the rise of respiration, but measurable amount are detected only after the onset of the climacteric.

Burg and Burg (1968) have challenged his assertion on the basis of following arguments:

Internal ethylene content and the rate of ethylene production are determined for a variety of fruit including several fruits previously been described as devoid of ethylene; e.g., the mango, pineapple, orange and lemon.

The application of ethylene hastens the ripening of mangoes even after the climacteric has began.

A question was raised as to whether or not ethylene is a true ripening hormone. They examined the role of propylene in respiration and ethylene production of tomato fruits of rin-a non ripening mutant. Propylene stimulated respiration in the immature fruits of rin, but there was no change in endogenous ethylene production. It was concluded that the onset of ripening in normal tomato fruit is not controlled by endogenous ethylene, and increased ethylene production is probably an integral part of the ripening process. This is the first work calling into question the fruit ripening hormone theory of ethylene.

Stimulation of Ripening

There are several aspects regarding the usefulness of stimulation of ripening. The most interesting consequence is the possibility of a once-over mechanical or manual harvest in crops that produce ripe fruits over a large time period and therefore normally must be harvest several times. A second aspect is the possibility to speed up the time of cultivation of a crop, especially in greenhouses where follow up crops are planted immediately. A further advantage of stimulating ripening is the possibility to plan harvest an marketing times of fruits. It some crops it may also be of interest, to harvest and market fruits generally earlier in the year.

Ethephon is used for stimulation of tomato fruit ripening. The fruit quality is not negatively affected compared with normally ripened tomatoes. The application of the growth regulator about 8 days before harvest the 20% colored fruits in the field allows a once-over mechanical harvest without loosing a substantial amount of unripe fruits. Tomato fruit ripening cannot be enhanced by ACC sufficiently. However, a derivative of ACC, N-Formyl-ACC, is very effective in stimulation of tomato fruit ripening.

Ethylene gas is widely used commercially for ripening a variety of climacteric fruits and decoloring non-climacteric citrus fruits. However, the use of ethylene-releasing compounds to effect this response is confined to relatively new crops and normally when they are used as a preharvest application. This is probably due to the ease with which ethylene gas can be applied to harvested fruit, and the dangers of applying aqueous solution to these disease-susceptible organs.

The major commercial application of ethephon for ripening fruits is in processing tomatoes. Ethephon treatment of tomato plants was shown to have adverse effects on the germinability of older seeds, indicating that there was no benefit of ethephon application to tomatoes destined solely for seed production.

The Role of Ethylene in Fruit Ripening

It is clear from the fact that externally applied ethylene can induce ripening in fruit, and that potentially stimulatory concentrations of the gas are present in fruit at the inception and during ripening, that the gas plays some fairly direct role in ripening. The precise nature of that role has still to be elucidated.

It is likely that in the climacteric fruits ethylene plays at least three distinct roles during ripening. It is a component of the system of opposing hormonal factors responsible for the initiation of the ripening process and the transition from growth to senescence. it is, in its autocatalytic production of supra-optimal concentration of the gas, a product of the ripening process. The large secondary production of ethylene triggered by the priming of the initiation mechanism may play a role in coordinating the other ripening changes.

The honeydew melon is an extreme of this type of behaviour in that : (1) the rise in ethylene production occurs over a long period of time, and (2) the level of ethylene required for initiation of the climacteric is relatively high. In other fruits (such as the banana) the increase in ethylene production may only precede the onset of the climacteric by a few hours. The other two types of relationship described invoke the presence of other factors inhibitory to ripening. These factors may either be endogenous products of the fruit or be produced outside the fruit and be continuously transported to the fruit from other parts of the parent plant. In the avocado the level of ethylene is not the factor that determines the onset of ripening, and that the loss of a factor supplied to the fruit while attached to the tree and disappearing slowly after harvest is the dominant feature of the initiation system. In the apple, attachment does not inhibit ripening, but it does modify the extent of the respiration climacteric and severely reduces ethylene production. In fruits such as bananas and apples, potentially stimulating levels of ethylene are frequently present in the fruit long before the onset of ripening. It appears, that in these cases, the decrease in another, as yet unidentified factor, by altering the sensitivity of the tissue of the ethylene already present, appears to determine the time of the initiation of ripening. In these fruits, the levels of exogenous ethylene required to promote ripening decrease as the fruit matures. It was shown that with very immature apples or bananas short ethylene treatments, which did not in themselves immediately promote ripening, shortened the time of the ventual onset of ripening compared with untreated controls. Thus, ethylene appears to affect the aging of the tissue and may act in promoting the disappearance of the other factors.

The concentration of ethylene in the interior of bulky organs is determined by the rate of production of the gas and by the structure of the tissue. In fruits a gaseous gradient of ethylene is established, which is due to the resistance to diffusion offered by the tissues and the relatively low surface area/volume of fruits. In fruits such as the apple, the epidermal layers offer the prinicpal barrier to diffusion. The development of stimulatory levels of ethylene inside the tissue can be an important part of the ripening initiation process. Changes in these levels can be dependent on changes in rate of production, but it is possible that changes in the internal ethylene levels.

It has been suggested that the deposition of wax on the skin of fruit and the clogging of the air spaces with solutes could lead to an increased resistances to diffusion, but these changes are only likely to affect the gaseous exchange in ripe fruit and would not be likely to interfere with changes involved in the initiation of the ripening process. The phenomenon of auto-inhibition of ethylene production has been suggested as a mechanism for the maintenance of the preclimacteric state.

In studies on figs (at the non-ripening stage) and on mature green tomatoes they found that if they allowed ethylene to accumulate around the fruit, ethylene production was stopped once a critical concentration had been reached (about 0.6 ppm), while no such auto-inhibition of ethylene production occurs if the experiment is repeated with figs in the ripening phase, or with the ripe tomatoes.

The second involvement of ethylene in ripening is the autocatalytic production of the gas; this can be considered as a ripening change alongside those of color, texture, and flavor. In many fruits, once ripening is initiated there is a rapid increase in the rate of production of ethylene and the internal concentration of the gas rises to levels as high as 1000 ppm. Initially, ethylene production increases logarithmically to about the time of the climacteric peak and then the rate of increase declines.

The internal concentrations can rise to several hundred ppm, a concentration well beyond the level required to saturate all known ethylene effects on plants, and the significance of this over-production is unclear.

It was found storage of uninitiated bananas in low oxygen atmospheres (1 to 5%) led to an almost complete inhibition of ripening, which could be reversed by application of ethylene in the low oxygen atmosphere. Similarly, in bananas in which ripening had been initiated by a short treatment with ethylene, ripening was delayed if the fruit was stored under low oxygen conditions (0 to 2%), even though the control of ripening was less than with uninitiated fruit. However, this effect could be reversed by including ethylene in the low oxygen atmosphere. This has led to the suggestion that ethylene has an important role in promoting directly the progress of some ripening changes, as well as in the initiation of ripening. There is evidence that ethylene may directly affect particular ripening processes and that activation of some events in ripening may be a direct response to the presence of high levels of ethylene. The best studied example is the relationship between ethylene treatment and the induction of the rise in respiration. A number of studies have shown that ethylene treatment of a wide range of plant tissues other than fruits can lead to a rise in respiration and this, of course, occurs without the attendant ripening changes. Among fruit tissues, including the nonclimacteric citrus fruits and the rin mutants of tomato, the response in respiration is initiated by ethylene treatment, but requires the continued presence of ethylene for the respiratory response to be sustained.

There are similar direct relationships between the presence of ethylene and other aspects of ripening. It has been suggested that the presence of high levels of ethylene has an effect on the softening of apples during ripening in store. These workers showed that the presence of an ethylene-absorbing compound in the store, which reduced the level of ethylene from about 1000ppm to below 10ppm, had a marked effect in retaining the texture of the fruit after storage.

There is some evidence that the threshold levels for the different ethylene-dependent responses of plant tissue may be different. In fruits there is also some evidence that the various processes of ripening are initiated at different internal concentrations of ethylene. It was shown that the climacteric in honeydew melons required a threshold of ethylene concentration of about 3 al/1 and a concentration for half-maximal response of 5 to 10 al/1. Thus different ripening processes seem to have different sensitivities to ethylene. it is thus tempting to suggest a role for ethylene in coordinating the sequence of ripening changes, as its concentration rose and individual responses were initiated. It must, however, be borne in mind that similar ripening changes occur in the non-climacteric fruit in the absence of increased ethylene production and the role of ethylene in the climacteric fruit may be to accelerate processes that proceed slowly in its absence.

Burg and Burg, in a study of the concentration dependence of a number of different responses of plant tissue to ethylene, found that the concentration for apparent half-maximal biological activity was 6 * 10-10 M. From a study of the relative activity of related hydrocarbons on the pea stem section assay, Burg and Burg defined the molecular requirements for biological activity. Burg and Burg proposed the presence of a metal binding site of ethylene. In view of the requirement of O_2 for ethylene action, they proposed that O_2 was necessary for ethylene to bind to the site. Since different ethylene responses show different concentration dependencies on the gas, there may be more than one ethylene binding site within the cell, each perhaps specific for a single response or group of responses.

Since the nature of the binding site is unknown, it is unclear whether ethylene acts in the aqueous or hydrophobic regions of the cell, but ethylene has an appreciable solubility in both water and lipids. In view of the gaseous nature of ethylene and the ease by which it is lost, it is perhaps surprising to find reports that ethylene gas is metabolized by tissues. CO_2, a known inhibitor of ethylene action, markedly inhibits $^{14}C_2H_4$ conversion to $^{14}CO_2$ without affecting the incorporation into soluble components.

Ethylene Genes and Fruit Ripening

It is now recognized that ethylene plays an essential role in the ripening of climacteric fruits. The cloning of the genes involved in ethylene biosynthesis during fruit ripening has permitted scientists to use biotechnology to genetically engineered crop plants with reduced endogenous ethylene production and fruits with prolonged storage life. It is possible to reduce ethylene production by lowering the expression of either ACC synthase or ACC oxides genes. Application of antisense RNA technology is such an example. In antisense RNA technique, a target gene (or cDNA) is inserted in the reverse orientation (antisense) to a plant through A. tumefaciens-mediated transformation. When both antisense and sense (parent) genes are expressed in the tissue, they tend to hybridize through sequence complementarily and the subsequent translational process is prevented. Such an approach has previously been undertaken with success to inhibit the expression of polygalacturonase gene in tomato fruit.

So far, efforts to alter ethylene biosynthesis in plant tissue have been focused on ripening fruits. Hamilton et al. (1990) were the first to reduce ethylene biosynthesis in tomato plants by transformation with an antisense pToM13 construct. Since ethylene production in the antisense-bearing plants was greatly reduced, they suggested that pToM13 may be related to ACC oxidase. Ripening of the transgenic fruis was retarded and their storage life extended. A similar approach was undertaken by Oeller et al. (1991) who transformed tomato plants with the antisense ACC synthase gene. Fruits from transgenic plans bearing expressed antisense RNA produced little ethylene and did not undergo ripening unless exogenous ethylene was provided. These results demonstrated that antisense technology can be successfully used to reduce ethylene production and thereby to extend the shelf life of fruits and vegetables. Alternatively, ethylene biosynthesis can be reduced by lowering the cellular level of ACC by introducing a bacterial gene encoding ACC deaminase into plants. Klee et al. (1991) cloned an ACC deaminase gene from pseudomonas species and expressed in tomato plants. Ethylene biosynthesis in the transformed plans was greatly reduced and ripening of fruits was similarly delayed. More recently, the ACC oxidase antisense gene has been used to delay the ripening of Cantaloupe melon (Ayub *et al.,* 1996). Antisense gene from heterologous source was also proven to be useful. Transgenic apple ACC oxidase antisense gene was shown to be able to reduce the ethylene biosynthesis in tomato and hence delay its ripening (Bolitho *et al.,* 1997).

The data obtained from the antisense ACC oxidase and ACC synthase fruits and results obtained from the expression of ACC deaminase in fruit proven that synthesis of ethylene during ripening occurs solely from the SAM cycle. The findings that the application of ethylene to the transgenic fruits can induce the climacteric respiratory rise in the absence of endogenous ethylene production and reverse the antisense phenotype demonstrates that ethylene controls the climacteric rise in respiration and is a causal agent of fruit ripening and not a simply a by-product of the ripening process. Other data supports the hypothesis that ethylene alone is not the sole determinants of ripening, since application of ethylene to transgenic ACC oxidase antisense fruit fail to cause complete reversal of the antisense phenotype. The visual difference observed in the phenotype of either ACC oxidase antisense or ACC deaminase fruits when left attached to the plant of detached and stored in air supports the involvement of a further factor associated with attachment of the fruit to the plant. The identify of this factor either a fruit associated inhibitor which is metabolized, mobilized or inactivated at the onset of ripening or an enhancer imported from the plant remains to be elucidated.

The pertinent Question is what mechanism is involved in the rapid decline of ethylene production that characterizes the unique ethylene biosynthesis in ripening. The evidence has suggested that ACC oxidase plays an important role in the regulation of ethylene production in fruits such as tomato, melon, banana, apple and pear. In has been established that ACC oxidase has an absolute requirement for ascorbate and iron as cofactors for its activity. Both ascorbate and iron contents increase temporarily at the onset of ripening and then decrease rapidly coinciding with the rapid decrease in the in vivo ACC oxidase activity. Liu et al. (1999) suggested that ACC oxidase decrease rapidly limitation of its co factors which in turn causes a rapid rise and fall of ethylene production at the early phase of ripening.

INHIBITORS AND FRUITSET

In Australia, Coombe (1985) treated four varieties of *Vitis cinifera* grape ("Zante' currant, 'Sultanina,' 'Muscat of Alexandria, and 'Doradillo') with CCC or Phosfon-D and found that when treatments were made two to three weeks before anthesis, the number of berries that set was considerably increased in all cultivars except 'Doradillo.' This was first time an increase in set of seeded berries ('Muscat of Alexandria') had been induced chemically in grape. In no instance was set increased significantly by treatment with these substances either at or after anthesis.

Pre-and postbloom application of SADH at a concentration of 500 ppm tend to increase fruit set in apple (McDonnell and Edgerton, 1970). However, treatment with CCC tends to reduce set.

INHIBITORS AND FRUIT GROWTH

Inhibitors, especially those in the plant growth retardant group, usually decrease the size of most fruits, including apple, grape, and squash.

Abscisic Acid

Little is known concerning the effects of this compound on fruit growth. Davis (1968) discovered that a poor correlation exists between growth of cotton bolls and ABA content. The control of Maturation and the initiation of ripening is thought to be due to the interaction and balance between opposing promotory and inhibitory factors. Ethylene is one important promoting factor, ABA appears to be another. Externally applied ABA will hasten ripening in a climacteric fruit such as the tomato, and will advance ripening in grapes if applied during stage II of the growth cycle.

Abscisic acid (ABA) has been shown to stimulate ethylene production in pre-climacteric apples and tomatoes. There have now been a number of studies of changes in the level of ABA in tissues during ripening. In a number of climacteric fruits, for instance the pear and the avocado, the level of ABA of constant during maturation, but rises during ripening. These rises in ABA accompany the rise in ethylene production during ripening. There was a threshold rise in free ABA level during ripening of detached avocado at 19 C to a maximum level of 700 μg/100 g fresh weight. The peak in ABA corresponded in time to the peaks of ethylene production and of the respiration, and preceded the softening of the tissues. A similar pattern of change was found in the bound ABA fraction, although the peak occurred about one day later than that in free ABA. The ratio of free/bound ABA was about 10:1 throughout ripening, and thus the increase in ABA must represent net synthesis rather than the release of the bound form.

Increase in ABA level during ripening occurs in a wide range of fruits of different ripening pattern. For instance, Rudnicki has shown a threshold increase in ABA levels in the strawberry, in which ripening occurs in the presence of little or no ethylene. Similarly, there is a rise in ABA in citrus fruits, which also ripen in the presence of only very low levels of ethylene. ABA level increases during ripening of some climacteric (peach) and non-climacteric (grape, cherry) fruit that show a three phase growth cycle. In the grape the ABA level rises steeply at version (onset of period III), while in the peach the ABA level increased just before and during period III, and this increase tends to precede the increased ethylene production associated with ripening. External treatments that delay or advance ripening also delay or advance the changes in ABA. In the pear, cold treatments that have been shown to promote ripening also advance the increase in ABA. It was known that ABA level in pears increased by up to fourfold during ripening. Similarly, in grapes Coombe and Hales found that treatment of the fruit during stage II of growth with 2-chloro-ethylphosphonic acid, which degrades on contact with the tissue to yield ethylene, promoted ripening and advanced the onset of the rise in ABA, while treatment with benzothiazole, 2-oxy acetic acid, which delayed ripening, also delayed the onset of increased ABA accumulation. These results clearly show on association between ABA and ripening, but they do not elucidate its role. A detailed study of the changes in ABA level during the growth of tomatoes of both normally ripening varieties and ones showing ripening defects. In all these varieties an increase in ABA level was found during the major period of growth of the fruit, reaching a peak shortly before the time of onset of the respiration climateric and increased ethylene production; this confirmed the results of an earlier study. The increase in ABA appears to be due to synthesis rather than release of the bound form, which only represented a small portion of the total ABA. The peak in ABA occurs in advance of the symptoms of ripening in all the strains and then declines. Throughout the period of ABA accumulation the ethylene level is low, and during the period of rapid rise in ethylene production in normal fruit the ABA level is falling. Clearly, the pattern of changes in ABA in the tomato is very different from that found in the other fruit discussed and further work needs to be done before the differences can be appreciated. The facts remain that in all fruits studied ABA is present in the unripe condition and can play a role in interacting with other factors in promoting ripening. In addition, rather like ethylene, ABA synthesis appears in some cases, to be stimulated by the initiation of the ripening process. The analogy with the role of ethylene in ripening can be taken a step further if one speculates that the increased tissue content of ABA may have a role in regulating some ripening changes once the ripening process has been initiated. It is interesting to point out that the increase in ABA cuts across the boundary of the climateric and non-climacteric fruits, and it may be that ABA is the dominant ripening promoter in some non-climacteric fruits.

RIPENING: EXTERNAL CONTROL

The use of external factors in an attempt to control the ripening process has important implications in the storage of fresh commodities prior to marketing, and this control of the ripening process must be achieved while minimizing the deleterious effect of water loss, microbial infection, and the development of physiological disorders storage.

Effect of Light

Light delays the loss of chlorophyll in detached leaf systems, but the pigment system involved in this light effect has not been determined with certainty. The loss of chlorophyll was enhanced by Red light in detached tomatoes. The accumulation of lycopene in tomato could also be induced by Red light and reversed with Far Red. ß-carotene, on the other hand, attained similar levels with Red, Far Red, and dark treatments. Phytochrome appears, therefore, to be involved in the control of the biosynthetic route leading to lycopene, but not of the one leading to . ß-carotene. Light is required for optimum fruit coluration. The light intensities required to saturate the lycopene system in tomato are quite low. Exposure to light is necessary also for anthocyanin biosynthesis in apple, grape, and other fruits.

Light intensity is also important. For full colour development in grapes, 54% of sunlight intensity was sufficient, but further reduction caused proportional decrease in anthocyanin levels down to and absence of anthocyanin in darkened clusters. The need for high light intensities points to mechanisms other that phytochrome in at least some fruits. Photosynthetic systems might be involved, since anthocyanin biosynthesis seems to be dependent upon carbohydrate availability, as it is in other tissues.

Effect of Temperature

Ripening occurs in only a relatively narrow range of temperatures. At temperature below a certain critical temperature, many fruits of tropical and sub-tropical origin develop a disorder known as chilling injury, the symptoms of which include an inhibition of ripening. Colour changes associated with ripening in the tomato are inhibited in storage at temperatures above 30° C. it is generally true that many fruits fail to ripen properly beyond 30° to 35° C and that in most cases this inhibition is reversible. Above about 30° C bananas fail to ripen normally, the pulp becoming soft and watery.

Gaseous Composition of Storage Atmosphere

The use of modified atmospheres to delay ripening stems was established by Kidd and West on English apples. They found that raising the CO_2 level and decreasing that of O_2 in the storage atmosphere delayed ripening, and such a modified atmosphere, coupled with the use of temperatures just above the limit for the onset of low temperature disorders, is the basis for the widely used method of controlled atmosphere storage that has proved so successful with a wide range of fruits.

Effect of Oxygen Tension

Low levels of O_2 (between 1 and 5%) have been shown by a large number of workers to delay the onset of ripening in a number of different fruits. It appears that O2 has its principal effect in interfering with the initiation process that determines the onset of ripening. The inclusion of low levels of ethylene in the low oxygen atmosphere reversed the inhibition of these ripening changes and ultimately normal ripe fruits were produced, showing that the ripening as such was not inhibited by low O_2. Thus, oxygen can act in system by affecting ethylene biosynthesis, ethylene action, and the sensitivity of the tissue to ethylene. In conclusion, it seems very likely that the principal effect of low oxygen atmospheres have two effects in promoting strong life. One is due to the inhibitory effect of low O_2 tensions on respiratory metabolism while the other is due to its effect on ethylene action and the initiation of ripening processes. The relative importance of these tow types of inhibitory action may vary from fruit to fruit.

Oxygen tensions greater than atmospheric tend to promote ethylene biosynthesis and consequently to promote ripening in a umber of fruits. This effect is significant in citrus fruits, which, if placed in atmospheres of 34 to 99% 02 will have a normal climacteric rise, and their production of ethylene is markedly increased.

Effect on Carbon dioxide Tension

Increasing CO_2 concentrations in the atmosphere around fruits delay the onset of the climacteric. Concentrations of CO_2 from 3 to 10% are generally effective in this, although higher CO_2 levels can, in some fruits (such as apples) lead to physiological disorders. The effect of CO_2 in delaying ripening has been related to its effect on ethylene action. Burg has proposed that CO_2 acts as a competitive inhibitor for ethylene action, presumably by competing with ethylene for the attachment site. CO_2 will also inhibit respiration in bulky organs. In contrast to this, CO_2 stimulates the respiration of lemons, but this is due to a stimulation of dark fixation of CO_2 and the resultant accumulation in the tissue of organic acids, which act as respiratory substrates.

The Effect of Atmospheric Pressure

Burg and Burg showed that the ripening of bananas was completely inhibited if the fruit was stored at one fifth atmospheric pressure in pure oxygen, so that the atmospheric tension of oxygen was maintained. The inclusion of small amounts of ethylene in the atmosphere overcame the inhibitory effect of the reduction in pressure. The fact that added ethylene reversed the inhibition showed that the removal of ethylene was the factor responsible for the delay in ripening.

In mature green tomatoes, storage at 19 cm Hg pressure at a partial pressure of oxygen of 0.20 did not delay ripening as compared to similar fruit stored under normal atmospheric conditions. However, storage under reduced pressures at lower O_2 tensions did delay ripening.

Chapter 10

POLYPRENOLS AND LEAF SENESCENCE

Polyprenols are long-chain isoprenoid lipids which form an intermediate class of substances filling up the gap between low molecular isoprenoids and high molecular natural rubber. The structure of plant polyprenols is shown in figure (5). There are two main types of long chain linear polymers constructed by plant organisms: the tri-trans poly-cis prenols and di-trans poly-cis prenols. It is assumed that they are formed either from tri-trans geranylgeranyl pyrophosphate or from di-trans farnesyl pyrophosphate respectively and isopentyl pyrophosphate by the action of a cis-prenyl transferase (Hemming, 1981). The di-trans polyprenols are either of the size of 6-9 isoprene residues or of the size of usually more than 14,15 isoprene residues. The tri-trans polyprenols have usually the size of 10,13 isoprene residues.

The long-chain polyprenols (and dolichols) reside in biological membranes. The size of these molecules exceeds the thickness of the biological membrane and therefore the accommodation of these molecules within membranes is a complex biological task. It has been shown in the case of dolichols that its incorporation into model lipi dmembrane led to destabilizing of the bilayer structure and the formation of cylindrical-hexagonal phase (Valterson *et al.,* 1985). The literature on the accommodation of polyprenols in membranes is still lacking but one can assume that similar type of accommodation may be valid in this case as well (Ranjan, 2001). The detailed studies on subcellular localization of polyprenols in leaves of plants are still missing. The data of Hemming's group (Hemming, 1969) have shown that most of the polyprenols in leaves occur in chloroplasts as well as in chromoplasts and the remaining is associated with particulate fractions of the cell. This was confirmed in the studies by Swiezewska et al. (1993). The biological role of polyprenols is still not known. So far the use of polyprenols has been made in biochemical laboratories studying the biosynthesis of sugars polymers and formation of glycoproteins.

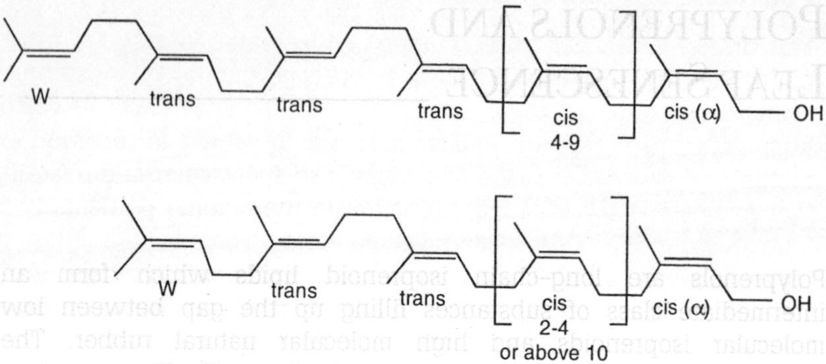

Fig. 10.1: Structure of plant polyprenols.

BIOSYNTHESIS OF POLYPRENOLS

The mechanism of biosynthesis of polyprenols has not been yet elucidated but it is believed to be the same as the formation of natural rubber. The early studies of Gough and Hemming (1970) demonstrated the biosynthesis of various long chain polyprenols from Mevalonic acid and in the work Swiezewska et al. (1993) mevalonate was found to be the precursor of prenol-10 and -11 in *in vivo* experiments with spinach leaves. However it is not yet clear whether the first starting compound is mevalonic acid giving rise to isopentenyl pyrophosphate or the non-mevalonate pathway described by Rohmer (Rohmer *et al.,* 1996) is operating. In the latter sequence the isopentenyl pyrophosphate is formed from deoxy-xylulose pyrophosphate. Since the long chain polyprenols are formed in the chloroplasts it seems reasonable to assume that the Rohmer's pathway is involved in their formation as being dominating in these subcellular structures. The chemical structure of polyprenols in which either two or three internal trans-isoprene units are located at the omega end of the molecule implies that farnesyl or geranylgeranyl fragment is the first precursor. Both of them are all-trans compounds. The elongation of these primers via the action of cis-prenyl transferase should lead to the expected poly-cis prenols. It is still not known in detail the mechanism of elongation and of termination of the polyprenol chain.

POLYPRENOLS AND AGING OF LEAVES

The disappearance of green colour in the leaves deciduous plants during autumn is one of the most impressive natural phenomena. It changes the appearance of the earth from space. A wave of colour from dark green through yellow and red to brown is recorded by orbiting satellites. The range of colours is so spectacular and beautiful that foliage is a major tourist attraction in the autumn particularly in Europe and in the North-Eastern United States. There is an increase of polyprenols content in the leaves of deciduous plants (Ranjan *et al.,* 2001) during autumn just before shedding of leaves. The increase of polyprenol content was not observed only in rhododendron species but also noticed in *Prunus, Sorbus* and *Rhus typhina* leaves. These results give hints that polyprenols are involved in the process of aging. High content of polyprenolsin yellow leaves in comparison to the content in green leaves in several similar phenomenon related to senescence and degreening of leaves has been observed in the case of tocopherol content in the leaves of *Melia azedarach* and *Vinca major*.

The presence of high amounts of polyprenols in yellow leaves raises a new aspect of future research concerning their possible physiological role in leaf senescence (Ranjan *et al.,* 2001).

Chapter 11

SENESCENCE IN VARIOUS CROP PLANTS

SENESCENCE PATTERN IN RICE (*ORIZA SATIVA* L.)

In monocarpic annuals the whole plant senescence generally occurs after the reproductive development and the cause of any change partioning to senescence becomes rather uncertain during this period (Davies *et al.,* 1977).

The rice plant seems to be a good experimental material for studying the mechanism of whole plant senescence in monocarpic annuals. The peculiar feature which characterizes this plant is that a special leaf, called the 'flag' leaf, emerges as the last leaf during the reproductive development, but the initiation of senescence of this leaf ensues much earlier than that of the two older leaves below it. Thus the typical sequential senescence observed in some other plants is quite lacking in rice and hence, this plant has been selected for the investigations.

Oryza sativa L. is of 135 days crop in which, seedling emerges within 5-6 day of sowing and tillering within 40-55 days. Ear emergence takes place at the plant age of 90 days and anthesis at 100 days, senescence starts at 130 days of sowing.

Experiment conducted out on senescence of rice shows. At the vegetative period (plant age 60 days), the cellular contents and the enzymic activities showed that the younger first leaf had a higher synthetic development it was observed that the flag leaf (youngest) was metabolically the most active while the third leaf the least. At the time of grain development (120 days), the above age-related metabolic activities of the three leaves completely changed and the flag leaf started senescing at a faster rate. The second leaf, however, remained physiobiochemically active even up to the time of grain maturation (130 days), when the flat and the third leaf partially senesced.

MECHANISM OF WHOLE PLANT SENESCENCE IN WHEAT (*TRITICUM AESTIVUM* L. CV. SONALIKA)

Monocarpic senescence in annuals has received much attention in recent years (Biswas and Choudhuri, 1980; Nooden, 1980a; Woolhouse, 1982a; Nooden, 1988c; Biswas and Ghosh, 1989; Ghosh, 1989; Ghosh and Biswas, 1989, 1991a, b).

A study of the senescence of sterile glumes of wheat may be a useful approach, since these are the photosynthetic organs closest to the grains and involved in the grain filling process (Thorne *et al.,* 1968). The extent of such involvement may vary among varieties (Saghir *et al.,* 1968; Lupton, 1969).

Triticum aestivum is hexaploid wheat with life period of 100-110 days which is grown in November and harvested in February. In wheat seedling emerges 5-6 day of sowing and ear emerges within 55-60 days. Grain develops between 60 days and yellowing of glumes starts at 90 days.

Experiment conducted out to study of the senescence of sterile glumes, flat leaf and the other two leaves below the ear of wheat (*Triticum aestivum* L. cv. Sonalika) shows that the senescence of sterile glumes was faster than of the leaves in terms of chlorophyll and protein degradation. The flag leaf senesced later than the other two leaves below it. Removal of sterile glumes markedly reduced the harvested index (crop : straw ratio) and average dry weight per grain as compared to removal of the flat leaf. Maximum grain weight was achieved after the glumes had senesced completely. Removal of the ear delayed senescence of all the tree leaves.

It is concluded that sterile glumes are important suppliers of assimilate for grain filling and that nutrient drainage is the primary cause of the monocarpic senescence in wheat.

Experiment conducted to the study of relative role of glumes, the flat leaf and the leaves below it in the correlative aspect of grain filling and metabolism rclated to the monocarpic senescence represents that flag leaf removal at any stage of grain growth hastened senescence (reduction in chlorophyll content) of the sterile glumes whereas a removal of the latter did not alter senescence of the former. Kernel mass, grain mass per ear, harvest index and sink activity reduced more conspicuously by the removal of glumes than by the flag leaf removal. Removal of grains hastened senescence of the glumes only, although protein content increased at a later grain removal in the majority of source organs. Moisture contents of the developing grains were mostly reduced by the removal of either the flag leaf or the glumes, though flag leaf removal at anthesis affected the most. Protein levels of the developing grains reduced at harvest only when the glumes were removed at anthesis, but sugar levels in later phases always decreased irrespective of the time of removal and nature of the source organs.

It is concluded that senescence of source organs (glumes and flag leaf) is directly related to grain filling of wheat.

IAA accelerated senescence (as determined by the loss of chlorophyll and protein) of the leaves and glumes (source) and reduced most of the physical yield parameters at harvest, although protein of the seeds and mobilization index for protein (grain protein: flat leaf plus glume protein ratio) increased by this treatment.

Application of kinetin or ABA either on the flag leaf or on the ear deferred senescence of the treated leaf or the glumes and also increased yield attributes excepting when its flag leaf was treated with kinetin. Proline levels were inversely related to the percentage of moisture levels of the seeds at harvest.

It is concluded that deferment of senescence of source and increment of yield of wheat are closely linked together.

MECHANISM OF MONOCARPIC SENESCENCE IN *PENNISETUM TYPHOIDES*

Pattern of monocarpic senescence of different annuals (both C_3 and C_4 plants) appears to be different. *Pennisetum typhoides* seems to be a good millet grain. C_4 annual plant for the study of monocarpic senescence, since in contrast to maize, where sinks (fruits) are situated at several nodal positions, the sink organ in *Pennisetum* is situated solely at the top (acrocarpous) and may, therefore, exert direct influence on the subtending source organ. Pearl millet

(*Pennisetum typhoides*) is cultivated widely in India for its edible and nutrition grain (July to October).

This plant as life cycle on 170 days in which seedling emerges. 5-6 days after seed sowing. Anthesis, grain formation and grain maturation starts at 80' 80-90 and 120-140 days respectively.

Though at anthesis the levels of chlorophyll and protein were lowest in the youngest flat leaf compared to the 3rd, 5th and 7th leaf (counted from the apex), the levels gradually increased in this particular leaf and showed the highest levels of these biomolecules at the time of completion of grain maturation (56th day after anthesis). The leaf thereafter senesced (as determined by the chlorophyll and protein loss) nonsequentially with the 3rd leaf (i.e. the flat leaf senesced earlier than the third leaf). Senescence was initiated in the 3rd, 5th and 7th leaf in a progressive manner from base to apex during the later stages of grain maturation. The levels of the above two cell components in the levels of the defruited plant was lower throughout the grain development period compared to their control counterparts. The leaves also senesced earlier, though the pattern of senescence remained unaltered. The third leaf of the defruited plant senesced abruptly when it subtended a daughter shoot.

Throughout the grain development period the soluble carbohydrate level was higher only in the third and fifth leaf, whereas reduced nitrogen level was always lower in the leaves of the defruited plant compared to those of the control. The levels of soluble carbohydrates and reduced nitrogen were higher in the stem of the defruited plant than those of the control. All the parameters were significantly decreased in the 3rd leaf when a daughter shoot developed from its axis. Fresh and dry weight of the grain reached their peaks within 56 days after anthesis. Compared to control, application of ABA as well as kinetin on the flat leaf increased the levels of chlorophyll protein soluble carbohydrates and reduced nitrogen of the said leaf and also increased the 1000 grain dry weight, grain yield per plant and soluble carbohydrate level of the grain at harvest, although reduced nitrogen levels in the grains decreased after 42 days following anthesis of the plant. Treatment of ABA on the spike produced almost similar results except that the levels of chlorophyll in the flag leaf decreased by the treatment. Application of kinetin on the spike reduced the levels of chlorophyll, protein and reduced nitrogen in the flat leaf, but increased 1000-grain weight and grain yield per plant at harvest. Soluble carbohydrate levels increased, but reduced nitrogen levels decreased in the grains of control and treated plants.

It is concluded that limitation in the capacity of accumulation of reduced nitrogen by the grains possibly offers increased self-sustenance to the leaf, which causes the deferment of leaf senescence in this species.

SENESCENCE PATTERN IN OIL CROPS

Mechanism of Monocarpic Senescence in Mustard

The life cycle of mustard is of 105 days. Flower initiation takes place after 45 days of sowing the seeds. Pod setting occurs between 55 to 70 days. The development of seeds occurs between 63 to 90 days and seeds get matured between 90 to days. In mustard senescence of leaves start before the onset of rapid pod growth and seed maturation. The exact cause of such senescence is still unknown. Nevertheless, some workers (Major and Charnetski, 1976; Major *et al.*, 1978) found that leaves of mustard contribute assimilate to the stem, pods and seeds before development of senescence. It is therefore apparent that flowering, storage of assimilate determines seed yield. However, according to Evans (1991) there is a complex complementary regulation of sink-source phenomenon in mustard at varying sink-source ratios.

Source-Sink Regulation During Seed Maturation By Developing Pods

In normal plants leaf senescence started and proceeded in a progressive manner from base to apex during the period of early pod setting. This suggests that pod walls and stems may act as temporary intermediary storage organs for the transfer of assimilates from leaves to seeds. Nevertheless, stem and pod walls have also some assimilatory function (Major and Charnetski, 1976)

According to (Biswas and Choudhuri, 1980; Noodén, 1980a, 1988b) there is a close parallelism between foliar senescence and maturation of fruits. However, In mustard it is quite different. In mustard in two different sets of acrocarpous plants containing 65 pods and 10 pods, respectively. Leaf senescence was delayed during the pod filling period. it started non-sequentially after complete yellowing and senescence of the pod wall. This indicates that seed filling is related to the senescence of leaf and also supports the nutrient diversion hypothesis of senescence and phenomenon of exhaustion death (Sinclair and Dewit, 1976)

There are reports that source-sink alterations do not affect photosynthesis of the source (Geiger, 1976, Noodén and Guiamet, 1989) but may affect the photosynthetic components (Noodén and Guiamet, 1989). However, according to Noodén and Guiamet, 1989 the reduced photosynthesis in fruitless plants may represent a metabolic adjustment, but not senescence. The increase in dry weight of the defruited mustard indicates that photosynthesis is geode up in this plants possibly due to presence of alternative sink in absence of fruits, photosynthesis in many plants often increased during early fruit growth when demand for carbohydrate and nitrogen is higher (Dornhalf and Shiples, 1970; Hanson *et al.*, 1976; Hall and Milthrope, 1978; Watson and Casper, 1984; Shibles *et al.*, 1987).

Removed of peduncle delays senescence of leaves over the removed of pedicel, but the senescence pattern remain unchanged. (non-sequential). The non-sequential mode of senescence can be explained as being due to reduction in the pod number to such extent that only the first leaf is drawn upon as a source leaf. Further more, the uppermost leaf of the early defruited plant (peduncles removed) senescence in the same way as the fruit-bearing acrocarpous 10-podded plant. This would suggest that a putative. Senescence stimulus is produced at the time of early floral differentiation (Choudhri and Mondal, 1980) and is perceived by the uppermost leaf already at this stage.

Source-Sink Regulation by Chemicals and Hormones

Among the chemically manipulated plants highest level of chlorophyll was found in the first leaf of urea-treated and lowest in the plants whose pods were injected (seed treatment) with kinetin, although in both the cases senescence of pod walls was, hastened dry weight of the plants and yield parameter were increased. Pod walls weight were almost parallel to those of seed weight but harvest indices decreased by these treatment.

This point out that nutrient remobilization out of the leaves to the pods cause quick fulfillment of storage requirement of the pod wall.

Senescence might hare resulted due to remobilization from leaf and quicker achievement of storage requirement in pod walls. There are reports that both cytokinin (Mothes *et al.,* 1959; Leopold and Kawase, 1964) and ABA (Milborrow, 1974; Dewdney and Mcwha, 1978; Bernier *et al.,* 1985) increase assimilate transport from leaves to fruits or untreated to treated organs.

Reduction in harvest indices by chemical treatments may due to limitation of the sink size (restriction of seed size) by the source capacity as well as greater retention of nutrients in the pod wall. So that it appears that there is a relation between nutritional status of the plant and pod wall senescence.

There results suggest that the pod wall serve as a temporary as well as intermediary storage organ and that foliar senescence is not directly to seed maturation in normal condition.

Mechanism of Monocarpic Senescence In Groundnut

The leguminous plants such as soybean and pea follows a pattern which favors the concept that migration of a 'Senescence signal' from the reproductive sink to the leaves (Malik and Berrie, 1975; Noodén, 1980a; 1988a,b) causes foliar senescence. Few plant also follow the pattern where nutrient withdrawal from the foliar organs causes monocarpic senescence. Grover *et al.,* 1985). According to Biswas and Mandal, 1993 both the hypothesis (nutrient drainage and senescence signal) might act in a super imposing manner and it is very difficult to represent them in a strict sense. Leguminous plants are indeterminate type where growth and development of vegetative and reproductive organs overlap each other.

Groundnut is a useful plant for the study of both 'nutrient drainage' and senescence signal' hypothesis, since cleistogamous (and some of chasmogamous) flowers situated below the leaves mature into fruits and act as strong sink. By contrast, aerial flowers do not mature into fruit and therefore may act only as a supplier of senescence signal.

The life cycle of groundnut is of 140 days. After 45 days of sowing the seeds, flower initiation takes place. Pod setting may start between 60-115 days. The period of seed development is approximately 90 to 130 days. The period of pod setting, seed development and seed maturation. Usually overlaps to each other in the case of groundnut. Yellowing of leaves start after senesce after 143 days of sowing the seeds. During harvesting a large quantity of foliage leaves remains green, which indicates that leaf senescence uncouples from pod maturation.

Source-Sink Regulation by the Developing Flowers And Fruits

In normal plant senescence parameter (level of chlorophyll and reduced nitrogen) increased up to the age of 101 days and after this decline.

In deflorated plant (fruit present) senescence parameter was highest compared to control and defruited plant. Therefore, it may be argued that aerial flowers might supply 'Senescence signal' to the subtending leaves leading to their senescence. According to Leopold et al (1959), removal of flowers is more effective than the young or ripening pods is dclaying senescence of leaves of soybean. This observation is consistent with the result obtainod in ground nut. Thus, the effect of nutrient withdrawal is less detrimental than senescence signal for the induction of whole plant senescence in groundnut, because the delay of leaf senescence by defruiting is less conspicuous than that by defloration.

Greater protein and reduced nitrogen accumulation the leaves of defruited plant might be associated with the non withdrawal of nutrients, while increase in sugar level in both the deflorated and defruited plant might be due to remobilization of assimilates However, the levels of protein and reduced nitrogen one not always directly related with the chlorophyll level of the leaf and hence senescence. However, Christensen et al. (1981), Crafts-Brandner et al. (1984a) obtained an inverse relation between the reduced nitrogen level in the leaf and the development of senescence in maize.

While 75% defoliation caused a decrease in all the parameters relating to senescence. However, 25% defoliation caused a slight decreased of chlorophyll level and aerial plant dry weight. However, the level of protein and reduced nitrogen in the leaf and sugar in stem increased. It suggests the compensatory role of existing leaves for the simultaneous reduction of sink size and increase in source capacity. However, (Williams et al. (1976), Boote et al. (1980) found a decrease in pod weight by different degrees of defoliation.

Increase of photosynthesis in intact leaf by partial defoliation plants are also found by (King, 1976; Carmi and Koller, 1979; Hanson and West, 1982).

Source-Sink Regulation By Hormones

Among the hormones kinetin or ABA treatment did not significantly changed the level of chlorophyll and aerial plant dry weight but the levels of protein and reduced nitrogen in leaf and sugar by kinetin. GA_3 increased almost all the parameters studied. All the three phytohormones increasing the sink size rather than sink pod by such treatment might be associated indicates that assimilates are retained in the pod wall.

Therefore, in groundnut the aerial flowers are more lethal than the underground fruits to induce whole plant senescence, where the leaf senescence is uncoupled from fruit maturation in normal condition possibly due to sink limitation.

Mechanism of Monocarpic Senescence In Til

The life cycle of Til is of 95 days. After 50 days of sowing the seeds flower initiation takes place. Pod setting occurs between 60 to 90 days. Seed development takes place between 75 to 85 days while seed maturation occurs between 85 to 95 days. The stages of seed development and pod setting overlaps to each other in the case of Til. When the seed is fully developed and matured senescence of pods prevails. The fruit of plant is a hairy capsule, developed at an axilliary position.

Source-Sink Regulation By Developing Capsule

In untreated control, capsule wall senescenced earlier than the subtending leaf. Defoliation hastened senescenced of capsule wall and also promoted seed filling. Although seed weight per capsule increased at harvest through defoliation, yield per plants deminished due to reduction in the number of capsular per plant. It is therefore, apparent that capsule walls of defoliated plant efficiently participated in seed filling. By increasing their own photosynthetic activity (in absence of leaf) or pre-flowering storage of assimilates helps in seed-filling. The capsule wall serves as intermediary organ for transport of pre-flowering assimilate accumulated in the stem. The increase in sugar level in capsule wall, inspite of inhibition of its own photosynthesis through darkening, also supports its role as an intermediary storage organ.

Defruiting and deflowering increased the chlorophyll and sugar levels of the subtending leaf. The order of increase in chlorophyll retention by the leaf amongst differently deflorated and defruited plant is:

Deflorated (complete defruiting) > defruited (flower present) > acrocarpous 10-capsulated (partial present) > unexcised control. The increase in chlorophyll and sugar level in the leaf due to defruting and deflowering may be due to increase in photosynthetic capacity of the leaf at the corresponding post-flowering stage and non-withdrawal of nutrients from the leaves to the fruits. But the reduction in dry weight in absence of reproductive structures may be due to concomitant increase in respiratory loss. However, there are evidenced where reproductive sink removal produced feed back inhibition of photosynthesis (Nosperger and Throne, 1965; King et al 1967, Hall and Brady, 1977, Hall and Milthrope, 1978) in leaf resulting from non-utilization of assimilates by the sink. Presence of flowers (defruited) produces more deleterious effect and defoliation naturally stopped capsule initiation, although seed weight per capsule (per-existing) significantly increased. This indicates that leaves serve for increase of the sink size (number of capsule per plant) without being directly involved as source for seed filling (capsule wall act as an intermediary supplier)

Source-Sink Regulation By Hormones

Treatment of kinetin on the capsule wall increased its chlorophyll, sugar and dry weight and also increased yield by increasing harvest index compared to seed treatment as well as control, although post-flowering contribution was reduced significantly by such treatment. Localized effect of kinetin in the fruit or leaf in increasing yield and delaying senescence, respectively, at the cost of remote organs, has been reported earlier (Mothes, 1961, Biswas and Mandal, 1986a, 1988b).

Treatment of ABA in the capsule wall defers senescence of both leaf and capsule wall and increased yield parameters. Seed treatment, though defers senescence of both leaf and capsule wall and increase dry weight of the plant. the seed weight per capsule decreases due to decrease in harvest index for capsule.

This indicates that translocation of sugar out of the capsule wall is inhibited inspite of increase in the level of sugar in the capsule wall compound to the leaves of such plant as well as leaves and capsule wall of controls. Never thless, increased translocation towards the ABA-treated seeds was observed by (McWha, 1975; Bernier et al 1982), where ABA stimulated sucrose transport (Brun et al 1986) Thus it is concluded that fruit maturation uncouples foliar senescence in normal condition and senescence in normal condition and senescence of leaf manipulated condition is more governed by nutrient exhaustion than others. Deferment of senescence of capsule wall along with increased translocation out of this organ at the post-capsule development period determines yield and in normal condition, contribution of source is controlled by sink limitation.

it occur almost at same time that is between 90 to 100 days. Yellowing of pods in storage rootless plants start after 95 days of sowing the seeds. while in rooted plants it start after 100 days.

SENESCENCE IN VEGETABLE CROPS

Mechanism of Monocaric Senescence in Radish

Radish seems to be an ideal system for studying the role of fruits as well as roots in the same or in different plants by manipulating the pattern of senescence, because in radish, some plants produce normal storage root, while others fails to develop the same even under non-competitive condition. Therefore, we can analysis the role of roots as well as pods for the development of monocarpic senescence in radish.

In two plants of radish in which one is with storage roots and 2nd is without storage roots, It is seen that the plants with storage roots are late maturing. Pod setting in plants 55 to 70 days while in plants with storage root, it occur between 70 to 80 days. Development of seeds in plants having no storage roots occur between 65 to 85 day and in rooted plants it occur between 70 to 90 days. Seed maturation in roots less plants take place between 85 to 100 days and in rooted plants,

Source-Sink Regulation by Storage Roots and Developing Pods

In radish having storage root, the first leaf from the top senesced earlier (non-seqential) than the 3rd leaf but the reverse (sequential) was found in plants devoid of storage root.

Such a differential response may be explained in terms of supply of cytokinin from the storage root to the leaves and nearest leaf is particular, for the deferment of senescence. The implication of root-formed cytokinin for the foliage has been widely discussed by (Sitton et al 1967; Kende, 1971; Noodén, 1988c; Noodén et al 1990).

Besides deferment of senescence of the first leaf, the reduction of pod load (acrocarpous depodded and acrocarpous) in plants with storage root. This also supports that root formed cytokinin defers senescence of the nearest leaf in the absence of reproductive sink.

Complete defoliation promoted the pod wall senescence in both types of plants and also reduced the seed weight per pod. This indicated that seed weight may be ascribed to the photosynthesis and storage of pod walls. This supports the earlier finding of Biswas and Mandal (1987b) in *Brassica* where pod wall functions as both assimilatory as well as intermediary storage organ for the transfer of assimilates from the leaf and stem.

Increase in whole plant dry weight through reduction in vegetative or generative sink by depodding of acrocarpous plant, decapitation and partial removal of storage root indicates that unexcised plant organs serve as alternate sink in the absence of appropriate sink strength and the storage root increased the source capacity by serving as an appropriate alternative sink. Thus proper source-sink adjustment is necessary for increased assimilation and growth at plant. This is consistent with the results obtained in other plants (Noodén and Guiamét, 1989).

Sink reduction through depodding of acrocarpous plant, decapitation and partial removal of storage root reduced the level of sugar also provide evidence for sink regulation of photosynthesis of source organ. At harvest, pod wall weight, seed weight and harvest index for storage organs were higher in the plants with storage root, whereas the HI for pod and seed were higher in plants without storage root. This indicates that accumulation exceed translocation in and out of the pod walls, causing earlier senescence of pod walls.

Source-Sink Regulation by Hormones

Application of kinetin on the developing pods delayed senescence of the first leaf in both the plants (with or without storage root). This result contradicts the result obtained in rice where application of kinetin to the spikelet promotes senescence of the flag leaf (Biswas and Mandal, 1986a).

However, application of kinetin on the pods promoted the senescence of pod walls in plants having storage root and delayed in plants without storage root.

This supports the earlier speculation (promotion of senescence through earlier fulfillment of storage capacity) of Biswas and Mandal (1987b) in case of mustard. In fact, the latter (treatment of kinetin on pod wall) is the localized effect of kinetin for withdraw of nutrients from the neighbouring source organs. (Mothes et al 1961; Leopald and Kawase, 1964) which seems to regulate senescence development.

It may therefore, be concluded that although senescence in most of the monocarpic annuals is triggered by flowering, the triggering effect is however, signaled by a coordinated control mechanism between source and sink i.e. accumulation exceeding redistribution as well as redistribution exceeding accumulation may produce senescence syndrome. In short, a proper balance between source and sink may be a regulatory factor of whole plant senescence.

Mechanism of Monocarpic Senescence in Potato

While most crop plants are annuals, not all fruit bearing annuals show monocarpic pattern of senescence in the strictest senesce (Woolhouse, 1982a), Most of potato plants under normal conditions die vegetatively after completion of tuberisation, But a few plant produce flowers and fruits which also die in a way similar to the plants which do not produce them at all. Therefore, this particular annual plant is not always monocarpic for the induction of senescence according to the definition of monocarpism. Also, there exists a source sink variation in this plant.

The life cycle of potato is of 90 days. Sprouting takes place after 10 days of sowing. The plant produces underground lateral runners known as stolons. The ends of stolon enlarge to form tubers. Vegetative plants of potato under normal condition die vegetatively after completion of tuberisation it means flowering does not occur in them. The initiation development and maturation of tubers take place between 35 to 85 days.

However, in some of the potato plants, which bear flower and fruits flower initiation, take peace 45 days of sowing. Fruit setting and development take place between 55 to 80 days. Yellowing of leaves start much earlier in these plants than that of vegetative plant.

Source-Sink Regulation by Reproductive Organs

In normal control plants the level of chlorophyll reached maximum up to the age of 55 days and after these declines.

The level of chlorophyll was higher in flower bearing plants than the fruit bearing, this indicates that the effect of nutrient withdrawal from the leaf (in fruit bearing plants) overrides the migration of senescence signal for the development of monocarpic senescence, because the leaf in the formed (flower bearing) senesces earlier than that of later (fruit bearing)

Nevertheless, the involvement of senescence signal, developing in the reproductive organs for the induction of leaf senescence may not be ruled out, since foliar senescence in flower-bearing plants proceeds at a much faster rate than in plants in the vegetative stage.

Dry weights of the plants, eliminated in reproductive maturation, were higher than those of vegetative conditions. This is in accordance with the finding that reproductive organ development promotes photosynthesis of the source organ (Dornhalf and Shibles, 1970) Rawson et al 1976; Lenz; 1977; Hall and Milthrope, 1978; Watson and Casper, 1984; Shibles et al 1987). In spite of the reduction in harvest indices, dry weight of tuber was higher in all plants bearing reproductive organs compared to those of controls. The inhibition of so mobilization out of the source is thus consistent with the increase in dry weight of such plant. Plants undergoing reproductive development require enhanced nutrient accumulation in response to appropriate photoperiod and temperature (Woolhouse 1982a).

Thus whole plant senescence in reproductive condition is therefore geared up by the factors others than nutrient remobilization alone. A tuber inducing substance, characteristically similar to jasmonic acid (Vrcugdentil and struik, 1989), may be implicated in such response as the chemical has a strong othylene generation capacity.

Source-Sink Regulation by Developing Tubers

However, the senescence of leaves started earlier in reproductive plants than the vegetative plants, but Higher chlorophyll level is in found leaves of detuperized vegetative plant. This supports that mechanism of senescence is related to remobilization of nutrients out of the leaf, as well as opined by several authors (Molisch, 1938; Biswas and Choudhuri, 1980; Ray and Choudhuri, 1981; Biswas and Mandal, 1986a, 1993).

Partial excision of roots decreased the level of chlorophyll in the leaf, Seemingly due to reduction in the supply of root-formed cytokinin (Itai and Vaadia, 1965; Skene, 1975; Van Staden and Davey, 1979; Carmi and Van Staden, 1983). The deferment of leaf senescence by foliar application of kinetin also supports this. Sharp reduction in tuber weight due to derooting along with the reduction in harvest index point out that translocation towards the tuber is greatly hampered. This is supported by the finding that sugar level was also highest in the leaf of such plant. Migration of root formed cytokinin towards the nearest sink (tuber) may be responsible for the creation of sink demand in normal condition.

Source-Sink Regulation by Hormones

CCC increased the chlorophyll level of the leaf, aerial plant weight, HI (harvest index) and tuber dry weight per plant, although post-tuberization contribution, decreased by such treatment. GA_3 reduced most of the parameters. This indicates that pre-tuberization contribution determines the increase in yield. Plant treated with kinetin and urea, delayed senescence of leaves and increased the economic yield due to 'balanced nutrient exhaustion out of the source organ. It is evident that, nutrient exhaustion towards the tuber cause whole plant senescence of potato in normal vegetative stage. But plants bearing reproductive organs, perhaps, experience other inductive phenomenon, in addition to nutrient exhaustion, for the development of monocarpic senescence. Therefore, the nature of monocarpic senescence (senescence due to reproductive development) and whole plant senescence (in normal vegetative plant) are different at least in this plant (potato).

Thus, the senescence in potato is therefore, not always coupled with the monocarpic nature of the plant, which seems to be genetically programmed developmental phenomenon. (Murfet, 1971; Proebsting et al 1977; King and Murfet 1985; Reid; 1987) The term monocarpic senescence should Therefore, be applied in a more restricted sense.

Mechanism of Monocarpic Senescence in *Amaranthus*

The life cycle of *Amaranthus* is of 140 days. Flower initiation takes place after 70 days of sowing the seeds. The inflorescence, developing on the extensive branches is a compound spike. The development and maturation of seed take place. Between 90 to 135 days. When seeds are completely developed senescence of leaves start.

There is large differences in the photosynthetic and yielding abilities between C_3 and C_4 plants, but relatively little is known about the various aspects of source-sink relationship during senescence of C_4 plants compared to the C_3 plant. The removal of major sink organs (flower, fruits etc) in C_3 plant generally defers senescence through nutrient retention, but the roverse is found in C_4 plants, The deferment or acceleration of senescence in C_4 annual like maize duc to removal of sink organs was, however, purported to be a variety character (Crafts-Brandner et al, 1984a). This indicate that source. Sink relationship is highly specific rather than uniform in C_4 annuals.

Source Sink Regulation by the Developing Flowers

In control plants, the chlorophyll level of leaves decreased after 125 days. In *Amaranthus* which is a C_4 plant removed of reproductive organs accelerates senescence of leaves, which is contrary to most of the C_3 plant. This may possibly be due to differences in metabolic activity of both types of plants. In balance in source : Sink ratio may also be the cause of such senescence. Increased sugar level in the vegetative organs such as leaf and stem of the deflorated plant at harvest also supports this. These results also speak against the possibility of a downward movement of senescence sign from the reproductive organ (Noodén, 1980a, 1988c) for the induction of senescence. Increase in the level of ABA after excision of reproductive organ (Loveys and Kriedemann, 1974; Setter et al 1980a, b; Noodén and Obermeyer, 1981) in this plant for the development of senescence may be a probability.

Contrary to the C_3 plants such as mustard (Biswas and Mandal, 1087b), the removed of reproductive organs reduces the growth of

Amaranthus by reducing the number of branches as well as dry weight of the plants.

Source-Sink Regulation By Hormones

In *Amaranthus* the application of kinetin on unexcised plant accelerates, although its application on deflorated plant defers senescence of leaves and also increases dry weight of the plant and the level of sugar in the stem at the later stage of growth. The results, therefore is contrary to the reports concerning the deferment of senescence of intact plant by exogenous application of kinetin (Mothes, 1959; Kulaeva, 1962; Biswas and Choudhri, 1980; Biswas and Mandal, 1988b). The results also indicate that the effect of kinetin on intact plant is apposite to that of deflorated plants. One of the reasons of the asymmetric responses of kinetin might be due to change of sensitivity of two types of plants to this hormone (Trewavas, 1981). This contrasting effect of kinetin was also reported by (Sexton and Woolhouse, 1984). The deferment of leaf senescence in deflorated plant by kinetin treatment may be due to the maintenance of proper nutritional balance by inducing the emergence of daughter shoot, concomitant with the diversion of sugar from the leaf to the stem.

However, it may be speculated that in intact *Amaranthus*, kinetin may produce mutual feed back interaction with the flowering and monocarpic senescence phenomenon are also supported to be genetically linked together. Thus, it may be concluded that a change of source-sink relation ship triggers the onset of senescence in *Amaranthus*.

SELECTED BIBLIOGRAPHY

1. Abadia, A., Poc, A. and Abadia, J. 1991. Could iron nutrition status be evaluated through photosynthetic pigment changes? *J. Plant. Nutri.* 14 : 987-999.
2. Altman A. 1982. Retardation of radish leaf senescence by polyamines. *Physiol. Plant.* 54, 189-193.
3. Ameisen, J.C. 1996. The origin of programmed cell death. *Science.* 272 : 1278-1279.
4. Amir-Shapira D, Goldschmidt EE, Altman A 1986. Autolysis of chlorophyll in aqueous and detergent suspensions of chloroplast fragments. *Plant Sci.* 43 : 201-206.
5. Anderson, J.M., Spilatro, S.R., Klauer, S.F. and Franceschi, V.R. 1989. Jasmonic acid-dependent increase in the level of vegetative storage proteins in soybean. *Plant Sci.* 62 : 45-52.
6. Apelbaum A, Icekson I, Burgoon A C and M. Lieberman 1982. Inhibition by polyamines of macromolecular synthesis and its implication for ethylene production and senescence processes. *Plant Physiol.* 70, 1221-1223.
7. Ayub, R., Guis, M., Amor, B., Gillot, L., Roustan, J.P., Latche, A., Bouzayen, M. and Pech, J.C. 1996. Expression of ACC oxidase antisense gene inhibits ripening of Cantaloupe melon fruits. *Nature-biotech.* 14 : 862-866.
8. Azumi, Y. and Watanabe, A. 1991. Eviodence for a senescence-associated gene induced by darkness. *Plant Physiol.* 95 : 577-583.
9. Bachmann A, Fernandez-Lopez J, Ginsburg S, Thomas H, Bouwkamp JC, Solomos T, Matile P 1994. Stay-green genotypes of *Phaseolus vulgaris* L: chloroplast proteins and chlorophyll catabolites during foliar senescence. *New Phytol.* 126: 593-600.
10. Bariol PA, Howard CJ, Taylor CB, Verburg MT, Jaglan VD, Green PJ. 1994. The limitation, Arabidopsis ribonuclease gene RNSI is tightly controlled in response to phosphate. *The Plant Journal.* 6, 673-85.

11. Bate NJ, Rothstein SJ, Thompson JE. 1991. Expression of nuclear and chloroplast photosynthesis-specific genes during leaf senescence. *Journal of Experimental Botany*. 42, 801-11.

12. Bazzaz, MB, Rebeiz, CA 1978. Chloroplast culture: the chlorophyll repair potential of mature chloroplasts incubated in a simple medium. *Biochem Biophys Acta*. 504 : 310-323.

13. Becker W, Apel K. 1993. Defferences in gene-expression between natural and artifically induced leaf senescence. *Planta*. 189, 74-9.

14. Belknap, W., and Garbarino, J. 1996. The role of ubiquitin in plant senescence and stress responses. *Trends in Plant Science*. 1:331-335.

15. Bernhard W, Matile P. 1994. Differential expression of glutamine synthetase genes during the senescence of *Arabidopsis thaliana* rosette leaves. *Plant Science*. 98, 7-14.

16. Beyer, E. M. Jr. 1976. A potent inhibitor of ethylene action in plans. *Plant Physiol*. 58 : 268-271.

17. Biswal, U.C., Kasemir H and Mohr H, 1982. Phytochrome control of degreening of attached cotyledons and primary leaves of mustard (*Sinapis alba* L.) seedlings. *Photochem. Photobiol*. 35: 237-241.

18. Biswal U.C. and Sharma R, 1976. Phytochrome regulation of senescence in detached barley leaves. *Z. pflanzenphysiol*. 80: 71-73.

19. Biswal, B., Choudhury, N K, Sahu P and Biswal U.C. 1993c. Senescence of detached fern leaves. *Plant Cell Physiol*. 24. 1203-1208.

20. Biswal, U.C., Bergfeld R and Kasemir R, 1993a. Phytochhrome-mediated delay of plastid senescence in mustard cotyledons: changes in pigment contents and ultrastructure. *Planta* 157: 85-90.

21. Biswal, U.C., Choudhury N. K. and Biswal B. 1983b. Degradation of plastid membranes during senescence. In *India Review of Life Science* (Edited by D.N. Sen and S. Kumar). Vol. 3. Pp. 33-61.

22. Bolitho, K.M., Lay-yee, M., Knighton, M.L. and Ross, G.S. 1997. Antisense apple ACC oxidase RNA reduces ethylene production in transgenic tomato fruit. *Plant Sci*. 122 : 91-99.

23. Borochov A, Woodson WR. 1989. Physiology and biochemistry of flower petal senescence. *Horticultural Review*. 11, 15-43.

24. Bortlik K 1990. Chlorophyllabbau: Charakterisierung von Kataboliten in seneszenten Gerstenblattern. Ph. D. Thesis, University of Zurich, Switzerland.

25. Bortlik K, Peisker C, Matile P. 1990. A novel type of chlorophyll catabolite in senescent barley leaves. *J. Plant Physiol*. 136: 161-165.

26. Bowen I.D. 1993. Apoptosis or programmed cell death? *Cell Biology International*. 17, 365-80.

27. Buchanan-Wollaston V, Ainsworth C. 1997. Leaf senescence in Brassica napus: cloning of senescence-related genes by subtractive hybrization. *Plant Molecular Biology* (in press).

28. Buchanan-Wollaston V. 1994. Isolation of cDNA clones for genes that are expressed during leaf senescence in Brassica napus. Plant Physiology 105, 839-46.

29. Buchanan-Wollaston, V. 1997. The molecular biology of leaf senescence. *J. Exp. Bot.* 48 : 181-199.

30. Burg, S.P. and Dijkman, M.J. 1967. Ethylene and auxin participation in pollen induced fading of vanda orchid blossoms. *Plant Physiol.* 61 : 812-815.

31. Butt VS, Lamb CJ. 1981. Oxygenases and the metabolism of plant products. In : Conn EE, ed. The biochemistry of plants, Vol. 7. New York: Academic Press, 626-65.

32. Callis J, Viestra RD, 1989. Ubiquitin and ubiquitin genes in higher plants. *Oxford Surveys of Plant Molecular Biology.* 6, 1-30.

33. Carfts-Brandner SJ. 1991. Non-structural carbohydrate metabolism during leaf ageing in tobacco (*Nicotiana tabacum*). *Physiologia Plantarum.* 82, 299-305.

34. Chen Q.G., Bleecker AB. 1995. Analysis of ethylene signal-transduction kinetice associated with seedling-growth response and chitinase induction in wild-type and mutant Arabidopsis. *Plant Physiology.* 108, 597-607.

35. Chen, S.J. and Kao, C.H. 1998. Methyl jasmonate, ammonium and leaf senescence in rice. *J. Plant Physiol.* 152 : 353-357.

36. Cheung AY, McNellis T, Piekos B 1993. Maintenance of chloroplast components during chromoplast differentiation in the tomato mutant green flesh. *Plant Physiol.* 101: 1223-1229.

37. Chubatsu L, Meneghini R. 1993. Metallotheionein protects DNA form oxidative damage. *Biochemical Journal.* 291, 193-8.

38. Ciechanover, A. 1994. The ubiquiten – proteasome proteolytic pathway. *Cell.* 79 : 13-21.

39. Clouse, S.D. 1996. Molecular genetic studies confirm the role of brasinosteroids in plant growth and development. *Plant. J.* 10 : 1-8.

40. Comai L, Dietrich RA, Maslyar DJ, Baden FS, Harada JJ. 1989. Coordinate expression of transcriptionally regulated isocitrate lyase and malate synthease genes in Brassica napus L. *The Plant Cell.* 1. 293-300.

41. Coupe SA, Taylor JF. Roberts JA. 1995. Characterization of an mRNA encoding a metallothionein-like protein that accumulates during ethylene-promoted abscission of Sambucus nigra L. leaflefts. *Planta.* 197, 442-7.

42. Csupor L 1971. Das Phytol in vergilbten Blattern. *Planta Med* 19: 37 – 41.

43. Curty C, Engel N 1996. Detection, isolation and structure elucidation of a chlorophyll *a* catabolite from autumnal senescent leaves of *Cercidiphyllum japonicum. Phytochemistry.* 42: 1531-1536.

44. Curty C, Engel N, Gossauer A 1995. Evidence for a monooxygenase-catalyzed primary process in the catabolism of chlorophyll. FEBS Lett 364: 41-44.

45. Dai, Y.R. and Galston A W 1981. Simultaneous phytochrome controlled promotion and inhibition of arginine decarboxylase activity in buds and epicotyls of etiolated peas. *Plant Physol.* 67: 266-269.

46. Davies KM, Grierson D. 1989. Identification of cDNA clones for tomato (Lycopersicon esculentum Mill) mRNAs that accumulate during fruit ripening and leaf senescence in response to ethylene. *Planta.* 179, 73-80.

47. De Greef, J A, Butler W L, Roth T.F and Frederieq H, 1971. Control of senescence in *Marchantia* by phytochrome. *Plant Physiol.* 48: 407-412.

48. Duggelin T, Bortlik K, gut H, Matile P, Thomas H 1988a. Leaf senescence in *Festuca pratensis*: accumulation lipofuscin-like compounds. *Physiol Plant.* 74: 131 –134.

49. Duggelin T, Schllenberg M, Bortlik K, Matile P 1988b. Vacuolar location of lipofuscin – and proline-like compounds in senescent barley leaves. *J Plant Physiol.* 133: 492-497.

50. Ecker JR. 1995. The ethylene signal transduction pathway in plants. *Science.* 268, 667-75.

51. Engel N, Jenny TA, Mooser V, Gosauer A 1991. Chlorophyll catabolism in *Chlorella protothecoids*: isolation and structure elucidation of a red bilin derivative. FEBS Lett 293: 131-133.

52. Engvild KC. 1989. The death hormone hypothesis. *Physiologia Plantarum.* 77, 282-5.

53. Evan-chen, Z, Mattoo A K and Goren R, 1982. Inhibition of ethylene biosynthesis by aminoethoxyvinyl-glycine and bypolyamines shunts label from 3,4-[^{14}C] methionine into spermidine in aged orange peel disc. *Plant Physiol.* 69: 385-388.

54. Feller U, Fischer A. 1994. Nitrogen metabolism in senescing leaves. *Critical Reviews in Plant Sciences.* 13(3), 241-73.

55. Feller U, Keist M. 1986. Senescence and nitrogen metabolism in annual plants. In: Lambers H, Neeterson J. J., Stulen I, eds. Fundamental, ecological and agricultural aspects of nitrogen metabolism in higher plants. Dordrecht: Martinus Nijhoff Publishers, 219-34.

56. Ferguson, D.L., A-Khatib, K., Guikema, J.A. and Ulsen, G.M. 1993. Degradation of proteins from the thylakoid membranes in senescing wheat leaves at high temperature. *Plant, Cell and Environ.* 16 : 421-428.

57. Foyer CH, Descourvieres P, Kunert KJ. 1994. Nucleotide-sequence and expression of a ripening and water stress-related cDNA from tomato with homology to the MIP class of membrane channel proteins. *Plant Molecular Biology*. 24, 539-43.

58. Fukuchi-Mizutani M, Savin K, Cornish E, Tanaka Y, Ashikari T. Kusumi T, Murata N. 1995. Senescence-induced expression of a homologue of delth 9 desaturase in rose petals. Plant Molecular Biology 29, 627-35.

59 Galston, A W, 1983. Polyamines as modulators of plant development. *Bioscience*, 33: 382-388.

60. Gan S, Amasino RM. 1995. Inhibition of leaf senescence by autoregulated production of cytokinin. Seience 270, 1966-7.

61. Gan, S. Adnm Amasino, R.M. 1997. Making sense of senescence. *Plant Physiol.* 113 : 313-319.

62. Garbarino J.E., Oosumi T, Belknap WR. 1995. Isolation of a polyubiquitin promoter and its expression in transgenic potato plants. *Plant Physiology*. 109, 1371-8.

63. Garcia A.L., Galindo L 1991. Chlorophyllase in *Citrus* leaves: localization and partial purification of the enzyme. *Photosynthetica*. 25: 105-111.

64. Genschik P, Durr A, Fleck J. 1996. Differential expression of several E2-type ubiquitin carrier protein genes at different developmental stages in *Arabidopsis thaliana* and *Nicotiana sylvestris*. *Molecular and General Genetics*. 244, 548-56.

65. Genschik, P. et al. 1994. Structure and promoter activity of a stress and developmentally regulated polyubiquitin encoding gene of *Nicotiar tabacum*. *Gene*. 148 : 195-202.

66. Gepstein S and Thimann K V 1980. Changes in the abscisic acid content of oat leaves during senescence. *Proc. Natl. Acad. Sci. USA* 77: 2050-2053.

67. Gepstein S., 1982. Light induced H^+ secretion and the relation to senescence of oat leaves. *Plant Physiol.* 70: 1120-1124.

68. Gepstein, S. and Thimann K. V, 1981. The role of ethylene in the senescence of oat leaves. *Plant Physiol.* 68: 349-354.

69. Gillissen, L.J.W. and Hoekstra, F. A. 1984. Pollination induced wilting in Petunia hybrida rapid transfer through the style of a wilting inducing substance. *Plant Physiol.* 74 : 496-498.

70. Ginsburg S., Matile P. 1993. Identification of catabolites of chlorophyll-porphyrin in senescent rapre cotyledons. *Plant Physiol.* 102:521– 527.

71. Ginsburg S., Schellenberg M., Maitle P. 1994. Cleavage of chlorophyll-porphyrin. Requirement for reduced ferredoxin an doxygen. *Plant Physiol.* 105: 545-554.

72. Goldthwaite J. J. and Laetsch W. M. 1967. Regulation of senescence in bean leaf discs by light and chemical growth regulators. *Plant Physiol.* 42: 1757-1762.

73. Goren R, Palavan N, Flores H, and Galston A W, 1982. Changes in polyamine titer in etiolated pea seedlings following red light treatment. *Plant Cell Physiol.* 23: 19-26.

74. Gosauer A 1994. Catabolism of tetrapyrroles. *Chimia.* 48: 352-361.

75. Gough, D. and Hemming, G. 1974. The stereochemistry of butalaprenoids biosynthesis. *Biochem.* J. 117 : 309-318.

76. Graham IA, Denby KJ, Leaver CJ. 1994. Carbon catabolite repression regulates glyoxylate cycle gene expression in cucumber. *The Plant Cell.* 6, 761-72.

77. Graham IA, Leaver CJ, Smith SM. 1992. Induction of malate synthase gene expression in senescent and detached organs of cucumber. The Plant Cell 4, 349-57.

78. Granell A, Harris N, Pisabarro AG, Carbonell J. 1992. Temporal and spatial expression of thiolprotease gene during pea ovary senescence and its regulation by gibberellin. *The Plant Journal.* 2(6), 907-15.

79. Grbic V, Bleecker AB. 1995. Ethylene regulates the timing of leaf senescence in *Arabidopsis. The Plant Journal.* 8(4), 595-602.

80. Green P.J. 1994. The ribonucleases of higher plants. *Annual Review of Plant Physiology and Plant Molecular Biology.* 45, 421-45.

81. Grierson D., Kader A. 1986. Fruit ripening and quality. In: Atherton J. Rudich J., eds. The tomato crop: a scientific basis for improvement. London: Chapman and Hall, 241-80.

82. Guiamet J.J., Schwatz E., Pichersky E., Nooden L.D. 1991. Characterization of cytoplasmic and nuclear mutations affecting chlorophyll and chlorophyll-binding proteins during senescence in soybean. *Plant Physiol.* 96: 227-231.

83. Gut H. Martile P. 1988. Apparent induction of key enzymes of the glyoxylic acid cycle in senescent barley leaves. *Planta.* 176, 548-50.

84. Haber, A H, Thompson P J, Waline P L and Triplett L L 1969. Non-photosynthetic retardation of chloroplast senescence by light. *Plant Physiol.* 44: 1619-1628.

85. Hamilton AJ, Lycett GW, Grierson D. 1990. Antisense gene that inhibits synthesis of the hormone ethylene in transgenic plants. *Nature.* 346, 284-7.

86. Hamilton, A.J., Lycett, G.W. and Grierson, D. 1990. Antisense gene that inhibits synthesis of the hormone ethylene in transgenic plants. *Nature.* 346 : 284-287.

87. Hanfrey C, Fife M, Buchanan-Wollaston V. 1996. Leaf senescence in *Brassica napus*: expression of genes encoding pathogenesis-related proteins. *Plant Molecular Biology.* 30: 597-609.

88. Hemming, F.W. 1969. Polyprenols. *Biochem. J.* 113 : 23-25.

89. Hemming, F.W. 1981. Biosynthesis of dolichol and related compounds. In: *Biosynthesis of isoprenoid compounds*, Porter, J.W. and Spurgein, S.L. eds. John Wiley and Sons, New York. 305-354.

90. Hendry G.A.F., Houghton J.D., Brown S.B. 1987. The degradation of chlorophyll: a biological enigma. *New Phytol.* 107: 255-302.

91. Hensel LL, Grbic V, Baumgarten DA, Bleecker AB. 1993. Developmental and age-related processes that influence the longevity and senescence of photosynthetic tissues in Arabidoposis. *The Plant Cell.* 5, 553-64.

92. Hinder B, Schellenberg M, Rodonis, Ginsburg S, Vogt E, Martinoia E, Matile P, Hortensteinder S 1996. How plants dispose of chlorophyll catabolites: transport of linear tetrapyrroles into isolated vacuoles. J. Biol. Chem. (in press).

93. Hinder, B., Schellenberg, M., Rodoni, S., Ginsburg, S., Vogt, E., Martinoia, E. et al. 1996. How plants dispose of chlorophyll catabolites. Directly energized uptake of tetrapyrrolic breakdown products into isolated vacuoles. *J. Biol. Chem.* 271, 27233-27236.

94. Hortensteiner S, Vicentini F, Matile P 1995. Chlorophyll break down in senescent leaves: enzymic cleavage of phaeophorbide a *in vitro. New Phytol.* 129: 237-246.

95. Hortensteiner, S. 1999. Chlorophyll breakdown in higher plants and algae. Cell Mol. Life Sci. 56 : 330-347.

96. Hsieh H.M., Liu W.K., Huang P.C. 1995. A novel stress-inducible metallothionein-like gene from rice. *Plant Molecular Biology.* 28, 381-9.

97. Huffaker R.C. 1990. Proteolytic activity during senescence of plants. *New Phytologist.* 116, 199-231.

98. Iordanov, I. T. and Merakchiiska-Niikolova M. G. 1983. Influence of darkening on the stability of plastid pigments and photosynthetic rate of primary bean leaves of different age of physiological state. *Photosynthetica*. 17:176-181.

99. Ito H., Takaichi S., Tsuji H., Tanaka A. 1994. Properties of synthesis of chlorophyll *a* from chlorophyll *b* in cucumber etioplasts. *J Biol Chem*. 269: 22034-22038.

100. Ito H, Tanaka A. 1996. Determination of the activity of chlorophyll *b* to chlorophyll *a* conversion during greening of etiolated cucumber cotyledons by using pyrochlorophyllide *b*. *Plant Physiol Biochem*. 34: 35-40.

101. Ito H., Tanaka Y., Tsuji H., Tanaka A. 1993 Conversion of chlorophyll *b* to chlorophyll *a* by isolated cucumber etioplasts. *Arch Biochem Biophys*. 306: 148-151.

102. Iturraspe J., Engel N., Gossauer A. 1994. Chlorophyll catabolism. Isolation and structure elucidation of chlorophyll *b* catabolites in *Chlorella protothecoides*. *Phytochemistry*. 35: 1387-1390.

103. Iturraspe J., Moyano N. 1995. A new 5-formylbilinoen as the major chlorophyll *a* catabolite in tree senescent leaves. *J Org Chem*. 60: 6664-6665.

104. John I., Drake R., Farrell A., Cooper W., Lee P., Horton P. Grierson D. 1995. Delayed leaf senescence in ethylene-deficient ACC-oxidase antisense tomato plants-molecular and physiological analysis. *The Plant Journal*. 7(3), 483-90.

105. Jones M.L., Larsen P.B., Woodson W.R. 1995. Ethylene-regulated expression of a carnation cysteine proteinase during flower petal senescence. *Plant Molecular Biology*. 28(3): 505-12.

106. Kamachi K. Yamaya T, Hayakawa T, Mae T, Ojima K. 1992. Changes in cytosolic glutamine synthetase polyeptide and its mRNA in a leaf blade or rice plants during natural senescence. *Plant Physiology*. 98: 1323-9.

107. Kamachi K. Yamaya T. Mae T, Ojjima K. 1991. A role for glutamine synthetase in the remobilization of leaf nitrogen during natural senescence in rice leaves. *Plant Physiology*. 96: 411-17.

108. Kasemir H and Mohr H, 1982. Coaction of three factors controlling chlorophyll and anthocyanin synthesis. *Planta*. 156: 282-288.

109. Kasemir H., 1983. Light control of chlorophyll accumulation in higher plants. In *Encyclopedia of Plant Physiology* (Edited by W. Shropshire, Jr and H. Mohr) Vol. 16. Pp. 662-686. Springer-Verlag, Berlin.

110. Kaur-sawhney R., Shin L. M., Flores H. E. and Galston A. W. 1982. Relation of polyamine synthesis and titer to aging and senescence of oat leaves. *Plant Physiol*. 69: 405-410.

111. Kawakami N., Watanabe A. 1988. Senescence-specific increase in cytosolic glutamine synthetase and its mRNA in radish cotyledons. *Plant Physiology.* 88: 1430-4.

112. Kegi J. 1991. Overview of metallothioneins. In: Riordan J. Vallee B, eds. *Methods in enzymology.* 613-26.

113. Kim D.J., Smith S. M. 1994a. Molecular cloning of cucumber phosphoenol pyruvate carboxykinase and developmental regulation of gene expression. *Plant Molecular Biology.* 26, 423-34.

114. Kim D.J., Smith S.M. 1994b. Expression of a singal gene encoding microbody NAD-malate dehydrogenase during glyoxysome and peroxisome development in cucumber. *Plant Molecular Biology.* 26: 1833-41.

115. King G.A., Davies K.M., Stewart R.J., Borst W.M. 1995. Similarities in gene-expression during the post-harvest-induced senescence of spears and natural foliar senescence of asparagus. *Plant Physiology.* 108, 125-8.

116. Klee, H. J., Hayford, M. B., Kretzmer, K. A., Barry, G. F. and Kishore, G. M. 1991. Control. Of ethylene synthesis by expression of a bacterial enzyme in transgenic tomato plants. *Plant Cell.* 3: 1187-1193.

117. Koiwai A., Matsuzaki T., Suzaki F., Kawashima N. 1981. Changes in total polar lipids and their fatty acid composition in tobacco leaves during growth and senescence. *Plant and Cell Physiology.* 22, 19059-65.

118. Koizumi M., Yamaguchi-Shinozaki K. Tsuji H. Shinozaki K. 1993. Structure and expression of two genes that encode distinct drought-inducible cysteine proteinases in *Arabidopsis thaliana.* Gene. 129: 175-82.

119. Kornberg H.L. Beevers H. 1957. The glyoxylate cycle as a stage in the conversion of fat to carbohydrate in castor beans. *Biochimica et Biophysica Acta.* 26. 531-7.

120. Krautler B., Jaun B., Amrein W., Bortlik K., Schellengerg M., Matile P. 1992. Breakdown of chlorophyll: constitution of a secoporphinoid catabolite isolated from senescent barley leaves. *Plant Physiol Biochem.* 30: 333-346.

121. Krautler B., Jaun B., Bortlik K., Schellenberg M., Maitle P. 1991. On the enigma of chlorophyll degradation: the constitution of a secoporphinoid catabolite. *Angew Chem Int Ed Engl.* 30: 1315-1318.

122. Lanahan M.B., Yen H.C. Giovannoni J.J. Klee H.J. 1994. The Never Ripe mutation blocks ethylene perception in tomato. *The Plant Cell.* 6: 521-30.

123. Langmeier M., Ginsburg S., Matile P. 1993. Chlorophyll break-down in senescent leaves: demonstration of Mg-dechelatase activity. *Physiol Plant.* 89: 347-353.

124. Leshem, Y. 1984. Interaction of cytokinins with lipid-associated oxy free radicals during senescence: a prospective mode of cytokinin action. *Can. J. Bot.* 62 : 2443-2449.

125. Lichtenthaler H. K., Burgstahler R., Buschmann C., Meir D., Prenzek U. and Schonthal A. 1982. Effect of high light and high light stress on composition, function and structure of the photosynthetic apparatus. In *Stress Effects on Photosynthesis.* (Edited by R. Marcelle). pp. 353-370. Dr. W. Junk Pubishers, The Hague.

126. Lincoln J.E., Fischer R.L. 1988. Regulation of gene expression by ethylene in wild-type and rin tomato fruit. *Plant Physiology.* 88, 370-4.

127. Liu, X., Shiomi, S., Nakatsuka, A., Kubo, Y, Nakamura, R. and Inaba, A. 1999. Characterization of ethylene biosynthesis associated with ripening in banana fruit. *Plant Physiol.* 121 : 1257-1265.

128. Lloyd, E. J. 1980. The effects of leaf age and senescence on the distribution of carbon in *Lolium temulentum. J. Exp Bot.* 31:1067-1079.

129. Lohman K.N., Gan S., John M.C., Amasino R.M. 1994. Molecular analysis of natural leaf senescence in *Arabidopsis thaliana. Physiologia Plantarum* 92: 322-8.

130. Lovell, P.H., Lovell, P.J. and Nichols, R. 1987. The importance of the stigma in flower senescence in petunia (*Petunia hybrida*). *Ann. Bot.* 60 : 41-47.

131. Malik N. S. A. and Thimann K. V. 1980. Metabolism of oat leaves during senescence VI. Changes in ATP levels. *Plant Physiol.* 65: 855-858.

132. Malik N.S.A. 1987. Senescence in oat leaves: changes in translatable mRNAs. *Physiologia Plantarum.* 70: 438-46.

133. Markides S.C., Goldthwaite J. 1981. Biochemical changes during bean leaf growth, maturity and senescence. Contents of DNA, polyribosomes, ribosomal RNA, protein and chlorophyll. *Journal of Experimental Botany.* 32: 725-35.

134. Matile P., Duggelin T., Schllenberg M., Rentsch D., Bortlik K., Peisker C., Thomas H. 1989. How and why is chlorophyll broken down in senescen tleaves? *Plant Physiol Biochem.* 27: 595-604.

135. Matile P., Ginsburg S., Schellenberg M., Thomas H. 1987. Catabolites of chlorophyll in senescent leaves. *J. Plant Physiol.* 129: 219-228.

136. Matile P., Ginsburg S., Schellenberg M., Thomas H. 1988. Catabolites of chlorophyll in senescing barley leaves are localized in the vacuoles of mesophyll cells. *Proc. Natl. Acad. Sci.* USA. 85: 9529-9532.

137. Matile P., Schellenberg M. 1996. The cleavage of phaeophorbide *a* is located in the envelope of barley gerontoplasts. *Plant Physiol Biochem.* 34: 55-59.

138. Matile P., Schellenberg M., Vicentini F. 1996. Localization of chlorophyllase in the chloroplast envelope. *Planta.* (in press).

139. Matile P. 1992. Chloroplast senescence. In: Baker NR, Thomas H, eds. *Crop Photosynthesis: Spatial and Temporal Determinants.* Amsterdam: Elsevier, 413-40.

140. Maunders M. J. and Brown S. B., 1983. The effect of light on chlorophyll loss in senescing leaves of sycamore *(Acer Pseudoplatanus L.). Plant.* 158: 309-311.

141. Maxie, E.C., Farnham, D.S., Mitchell, F.G., Sommer, N.F., Pearson, R.A., Snyder, R.G. and Rae, H.L. 1973. Temperature and ethylene effects on cut flowers of carnation (*Dianthus caryophyllus* L.). *J. Am Soc. Hort. Sci.* 98 : 568-572.

142. McLaughlin J.C., Smith S.M. 1994. Metabolic regulation of glyoxylate-cycle enzyme synthesis in detached cucumber cotyledons and protoplasts. *Planta.* 195: 22-8.

143. McLaughlin J.C., Smith S.M. 1995. Glyoxylate cycle enzyme synthesis during the irreversible phase of senescence of cucumber cotyledons. *Journal of Plant Physiology.* 146, 133-8.

144. Meier D. and Lichtenthaler H. K., 1983. Ultrastructure of chloroplasts from wheat and maize plants grown at high and low qunata fluence rates. In *Photosynthesis and Plant Productivity.* (Edited by H. Metzner). pp. 2356-238. Wissenschftl. Verlags-gesellschaft. Stuttgart.

145. Merzlyak M.N., Hendry G.A.F. 1994. Free radical metabolism, pigment degradation and lipid peroxidation in leaves during senescence. *Proceedlings of the Royal Society of Endinburgh.* 102B: 459-71.

146. Mistuhashi W., Feller U. 1992. Effects of high and external solutes on the catabolism of nuclear-encoded stromal proteins in intact chloroplasts isolated from pea leaves. *Plant Physiology.* 100: 2100-5.

147. Mittler R. Lam E. 1995. Identification, characterization, and purification of a tobacco endonuclease activity induced upon hypersensitive response cell death. *The Plant Cell.* 7, 1951-62.

148. Mohr H. and Shropshire W. Jr, 1983. An introduction to phtomorphogenesis for the general reader. In *Encyclopedia of Plant Physiology* (Edited by W. Shropshire, Jr. and H. Mohr). Vol. 16. pp. 24-38. Springer-Verlag. Berlin.

149. Morison, J. I. L. and Jarvis P. G. 1983. Direct and indirect effects of light on stomata. 1. In scotpines and sitka spruce. *Plant Cell Environ.* 6: 95-101.

150. Muhitch, M. J., Edwards L. A. and Fletcher J S 1983. Infuence of diamines of polyamines on the senescence of plant suspension cultures. *Plant Cell. Rep.* 2: 82-84.

151. Muhlecker W., Hortensteiner S., Ongania K.H., Matile P., Krautler B. 199. A fluorescing chlorophyll catabolite from higher plans. *Chimia.* 50: 346.

152. Muhlecker W., Krautler B., Ginsburg S., Matile P. 1993. Break-down of chlorophyll: a tetrapyrrolic chlorophyll catabolite from senescent rape leaves. *Helv Chim Acta.* 76: 2976-2980.

153. Muller-Uri, F., Parthier, B. and Nover, L. 1988. Jasmonate-induced alteration of gene expression in barely leaf segment analyzed by in-vivo and in-vitro protein synthesis. *Planta.* 176 : 241-247.

154. Nakamura H., Kishi Y., Shimomura O., Morse D., Hastings J.W. 1989. Structure of dinoflagellate luciferin and its enzymatic air-oxidation products. *J Am Chem Soc.* 111: 7607-7611.

155. Nakamura H., Musicki B., Kishi Y. 1988. Structure of the light emitter in krill (*Euphausia pacifica*) bioluminescence. *J. Am. Chem. Soc.* 110: 2683-2685.

156. Nejidat A., Itai C. and Bejerano N. R. 1983. Stomatal response to ATP mediated by phytochrome. *Physiol. Plant.* 57: 367-370.

157. Nichols, R., Buffer, G., Mor, Y., Fujino, D.W. and Reid, M.S. 1983. Changes in ethylene and 1-amino-cylopropane-1-carboxylic acid content of pollinated canation flowers. J. Plant Growth Regul, 21 : 8.

158. Nooden LD, Singh. S. Letham DS, 1990. Correlation of xylem sap levels with monocarpic senescence in soybean. Plant Physiology 93, 33-9.

159. Noodén L.D. 1988. The phenomena of senescence and aging. In: Nooden LD, Leopold AC, eds. *Senescence and Aging in Plants.* San Diego: Academic Press.

160. Nurnberger, T. Abel, S., Jost, W. and Glund, K. 1990. Induction of an extracellular ribonuclease in cultured tobacco cells upon phosphate starvation. *Plant Physiol.* 93 : 33-39.

161. Oeller, P.W., Li, M.W., Taylor, L.P., Pike, D.A., and Theologis, A. 1991. Reversible inhibition of tomato fruit senescence by antisense RNA. *Science* 254 : 437-439.

162. Oh, S.A., Lee, S.Y., Chung, I.K., Lee, C-H. and Nam, H.G. 1996. A senescence-associated gene of Arabidopsis thaliana is distinctively regulated during natural and artificially induced leaf senescence. *Plant Mol. Biol.* 30: 739-754.

163. Oshio Y., Hase E. 1969. Studies on red pigments excreted by cells of *Chlorella protothecoides* during the process of bleaching induced by glucose or acetate. I. Chemical properties of the red pigments. *Plant Cell Physiol.* 10: 51-59.

164. Panigrahi P. K. and Biswal U. C., 1983. Aging of chloroplasts *in vitro*-III. Comparison of the effects of aging *in vitro* and senescence on the red absorption band and primary photochemical reaction of sunflower chloroplast. *Photosynthesis Res.* 4: 183-189.

165. Park K.Y., Drory A., Woodson W.R. 1992. Molecular cloning of an 1-aminocyclopropane-1-carboxylate synthase from senescing carnation flower petals. *Plant Molecular Biology.* 18: 377-86.

166. Park Y., Morris M.M., Mackinney G. 1973. ON chlorophyll break down in senescent leaves. *J. Agric. Food. Chem.* 21: 279-281.

167. Pastori G.M., del Rio L.A. 1994. An activated-oxygen-mediated role for peroxisomes in the mechanism of senescence of *Pisum sativam* L. leaves. *Planta.* 193, 385-91.

168. Pastori, G.M. and Del Rio, L.A. 1997. Natural senescence of pea leaves: An activated-oxygen mediated function for peroxisomes. *Plant Physiol.* 113: 411-418.

169. Peisker C. 1991. Chlorophyllabbau in vergilbenden Gerstenblattern: Entwicklung und Chlorophyll. Ph.D. Thesis, University of Zurich, Switzerland.

170. Peisker C., Thomas H., Matile P. 1990. Radiolabelling of chlorophyll for studies on catabolism. *J. Plant Physiol.* 136: 544-549.

171. Peisker, C., Duggelin T., Rentsch D., Matile P., 1989. Phytol and the breakdown of chlorophyll in senescent leaves. *J. Plant. Physiol.* 135: 428-432.

172. Pennazio, S. 1981. Changes in the soluble protein constitution of Gomphrena globosa leaves showing spontaneous local lesions. *Rivista di Patologia Vegetale.* 17 : 127-135.

173. Pfeiffer H. and Kleudgen H. K., 1980. Investigations on the phytochrome control of senescence in the photosynthetic apparatus of *Hordeum Vulgare* L *Z. Pflanzenphysiol.* 100: 437-445.

174. Picton S., Barton S.L., Bouzayen M., Hamilton A.J., Grierson D. 1993. Altered fruit ripening and leaf senescence in tomatoes expressing an antisense ethylene-forming enzyme transgene. *Plant Journal.* 3: 469-81.

175. Pilbeam, C. J. 1992. Effect of nitrogen supply on the growth and senescence of leaves of *Lilium perenne* with contrasting rates of leaf respiration. *Ann. Bot.* 70 : 365-370.

176. Pjon C. J. 1981. Effects of cyclohexamide and light on leaf senescence in maize and hydrangea. *Plant Cell Physiol.* 22: 847-854.

177. Pogson B.J., Downs C.G., Davies K.M. 1995a. Differential expression of two 1-aminocyclopropane-1-carboxylic acid oxidase genes in broccoli after harvest. *Plant Physiology.* 108, 651-7.

178. Pogson B.J., Downs C.G.,. Davies K.M., Morris S.C., Buchanan-Wollaston V. 1995b. Nucleotide sequence of a cDNA clone from broccoli with high identity with the PSST aubunit of NADH : ubiquinone oxidoreductase. *Plant Physiology.* 108, 859-60.

179. Ranjan, R., Chojnacki, T. and Swiezewska, E. 2001. On the occurrence of polyprenols in plants. *Polish J. Bot.* (accepted for publication).

180. Ranjan, R., Marczewski, A., Chojnacki, T., Hertal, J. and Swiezewska, E. 2001. Polyprenols and the phenomenon of the senescence in plants. *Acta Biochim. Polon.* (accepted for publication).

181. Rao G. N. 1981. Delaying of petal senescence in excised flowers of Dombeya by certain plant growth regulators. *Ind J Expp Biol.* 21: 47-48.

182. Raval M. K., Mishra A N and Biswal U. C. 1982. Changes in the membranes of wheat chloroplasts aging *in vitro*. *Photobiochem Photobiophys.* 4: 293-297.

183. Rechsteiner M. 1991. Natural substartes of the ubiquitin proteolytic pathway *Cell.* 66, 615-18.

184. Reid, M.S. and Wu, M-J. 1992. Ethylene and flower senescence. *Plant Growth Regul.* 11 : 37-43.

185. Reinbothe S., Mollenhauer B., Reinbothe C. 1994. JIPs and RIPs : the regulation of plant gene expression by jasmonates in response to environmental cues and pathogens. *The Plant Cell.* 6: 1197-1209.

186. Rhie G., Beale S.I. 1992. Biosynthesis of phycobilins. Ferrdoxin-supported NADPH-independent oxygenase and phycobilin-forming activities from *Cyanidium caldarium*. *J. Biol. Chem.* 267: 16088-16093.

187. Rohmer, M., Seeman, M., Horbach, S., Bringer-Meyer, S. and Sahm, H. 1996. Glyceraldehyde 3-phosphate and pyruvate as precursors of isoprenic units in an alternative non-mevalonate pathway for terpenoid biosynthesis. *J. Am. Chem. Soc.* 118 : 2564-2566.

188. Rohwer, F and Scheirle J j1982. Efect of light on ethylene production red light enhancement of 1-amino cyclopropane-1-carboxylic acid concentration in etiolated pea shoots. *Z. Pflanzenphysiol.* 107:295-300.

189. Rousseaux, M.C., Hall, A.J. and Sanchez, R.A. 1996. Farred enrichment and photosynthetically active radiation level influence leaf senescence in field growth sunflower. *Physiol Plant.* 96 : 217-224.

190. Ryu S.B., Wang X.M. 1995. Expression of phospholipase-D during castor bean senescence. *Plant Physiology*. 108, 713-19.

191. Satler S. O. and Thimann K. V. 1983. Metabolism of oat leaves during senescence VII. The interaction of carbon dioxide and other atmospheric gases with light in controlling chlorophyll loss and senescence. *Plant physiol.* 71: 67-70.

192. Satoh, H., Nakayama, K. and Okada, M. 1998. Molecular cloning and functional expression of a water soluble chlorophyll molecules in cauliflower. *J. Biol. Chem.* 273 : 30568-30575.

193. Scheffner, M., Nuber, U. and Huibregtse, J.M. 1995. Protein ubiquitination involving and E1-E2-E3 enzyme ubiquitin thioester cascade. *Nature*. 373 : 81-83.

194. Schellenberg M., Matile P., Thomas H. 1993. Production of apresumptive chlorophyll catabolite in vitro: requirement for reduced ferredoxin. *Planta*. 191: 417-420.

195. Schellenberg M., Matile P., Thoms H. 1990. Breakdown of chlorophyll in chloroplasts of senescent barley leaves depends on ATP. *J. Plant Physiol.* 136: 564-568.

196. Schmidt A, Jager K. 1992. Open questions about sulfur metabolism in plants. Annual Review of Plant Physiology and Plant Molecular Biology 43, 325-49.

197. Schmutz D, Wyss H R and Brunold C 9183. Activity of sulfate-assimilating enzymes in primary leaves of *Phaseolus vulgaris L.* during dark induced senescence *Z. Pflanzenphysiol.* 110: 211-220.

198. Shankin J., DeWitt N.D., Flanagan J.M. 1995. The stroma of higher plant plastids contain Clpp and ClpC, functional homologs of *Escherichia coli* Clpp and ClpA: an archetypal two-component ATP-dependent protease. *The Plant Cell.* 7, 1713-22.

199. Shih, L. M., Sawhney R. K., Fuhrer J., Samanta S. and Glaston A. W., 1982. Effects of exogenous 1,3-Diamino propane and spermidine on senescence of oat leaves I. Inhibition of protease activity, ethlene production and chlorophyll loss as related to polyaminc content. *Plant Physiol.* 70: 1592-1596.

200. Shioi Y., Sasa T. 1986. Purification of solubilized chlorophyllase from *Chlorella protothecoides. Methods Enzymol.* 123: 421-427.

201. Shioi Y., Tatsumi Y., Shimokawa K. 1991. Enzymatic degradation of chlorophyll in *Chenopodium album. Plant Cell Physiol.* 32: 87-93.

202. Shioi Y., Tomita N., Tsuchiya T., Takamiya K., 1996. Conversion of chlorophyllide to pheeophorbide by Mg-dechelating substance in extracts of *Chenopodium album. Plant Physiol. Biochem.* 34: 41-47.

203. Singh, S., Letham, D.S. and Palni, L.M.S. 1992. Effect of growth substances and mineral nutrients on cytokinin levels and senescence of tobacco (*Nicotiana rustica*) leaves. In : Karssen, C.M., van loon, L.C., Vreugdenhil, D. eds. *Progress in Plant Growth Regulation.* Dordrecht: Kluwer Academic Publishers, 628-634.

204. Smart C.M. 1994. Gene experssion during leaf senescence. *New Phytol.* 126: 419-448.

205. Smart C.M., Hosken S.E., Thomas H., Greaves J.A., Blair B.C., Schuch W., 1995. The timing of maize leaf senescence and characterization of senescence-related cDNAs. *Physiol Plant.* 93: 673-682.

206. Smart C.M., Hosken S.E., Thomas H., Greaves J.A., Blair B.G., Suhuch W. 1995. The timing of maize leaf senescence and characterization of senescence-related cDNAs. *Physiologia Plantarum.* 93, 673-82.

207. Smart C.M., Scofield S.R., Bevan M.W., Dyer T.A. 1991. Delayed leaf senescence in tobacco plants transformed with tmr, a gene for cytokinin production in Agrobacterium. *The Plant Cell.* 3, 647-56.

208. Smart C.M. 1994. Gene expression during leaf senescence. *New Phytology.* 126, 419-48.

209. Smart, C. M. 1994. Gene expression during leaf senescence. *New Phytol.* 126 : 419-448.

210. Steintz B., Cohen A. and Leshem B., 1980. Factors controlling the retardation of Chlorophyll degradation during senescence of detached statice (*Limonium sinuatium)* flower stalks. *Z. Pflanzenphysiol.* 100: 343-349.

211. Stoddart J. L. and Thomas H., 1982. Leaf senescence. In *Encyclopedia of Plant Physiology* (Edited D. Boutler and B. Parthier), Vol. 14A. pp. 592-636. Springer-Verlag. Berlin.

212. Stoddart L., Thomas H. 1982. Leaf senescence. In: *Encyclopedia of Plant Physiology*, New series, Vol. 14A. Springer Verlag, 592-636.

213. Swiezewska, E., Dallner, G., Andersson, B. and Ernster, L. 1993. Biosynthesis of ubiquinone and Plastoquinone in the endoplasmic reticulum-Golgi membranes of spinach leaves. *J. Biol Chem.* 268 : 1494-1499.

214. Szamosi I., Shaner D.L., Singh B.K. 1993. Identification and characterization of a biodegradative form of threonine dehydratase in senescing tomato (Lycopersicon esculentun) leaf. *Plant Physiology.* 101, 999-1004.

215. Taylor C.B., Bariola P.A., Delcardayre S.B., Raines R.T., Green P.J. 1993. RNS2-a senescence-associated RNAse of Arabidopsis that diverged from the sRNAses before speciation. *Proceedings of the National Academy of Sciences* USA. 90: 5118-22.

216. Teramoto H., Toyama T., Takeba G, Tsuji H. 1995. Changes in expression of cytokinin-repressed genes, CR9 and CR20, in relation to aging, greening and wounding in cucumber. *Planta.* 196, 387-95.

217. Theil E.C. 1987. Ferritin: structure, gene regulation, and cellular function in animals, plants and microoragnisms. *Ann. Rev. Bioch.* 56: 289-315.

218. Theologis A. 1993. One rotten apple spoile the whole bushel: the role of ethylene in fruit ripening. *Cell.* 70, 181-4.

219. Thimann K. V. and Satler S. O. 1979. Relation between leaf senescence and stomatal closure: senescence in light. *Proc. Natl. Acad Sci USA.* 76: 2295-2298.

220. Thimann K.V. 1980. The senescence of leaves. In: Thimann KV, ed. *Senescence in Plants.* Boca Raton: CRC Press, 85-115.

221. Thomas H., Bortlik K., Rentsch D., Schellenberg M., Matile P. 1989. Catabolism of chlorophyll in vivo: significance of polar chlorophyll catabolites in a non-yellowing senescence mutant of *Festuca pratensis* Huds. New Phytol 111: 3-8.

222. Thomas H, Hilditch P 1987. Metabolism of thylakoid membrane proteins during foliar senescence. *In* WW Thompson, EA Nothnagel, RC Huffaker eds. *Plant Senescence: Its Biochemistry and Physiology. The American Society of Plant Physiologists,* Rock-ville, MD, pp. 114-122.

223. Thomas H., Schellenberg M., Vicentini F., Matile P. 1996. Gregor Mendel's green and yellow pea seeds. *Bot Acta.* 190: 3-4.

224. Thomas H, Smart CM. 1993. Crops that stay green. *Ann. App. Biol.* 123: 193-219.

225. Thomas H., Smat C.M. 1993. Crops that stay green. *Ann Appl Biol.* 123: 193-219.

226. Thomas, H., Ougham H.J., Davies T.G.E. 1992. Leaf senescence in a non-yellowing mutant of Festuca partensis-transcripts and translation products. *Journal of Plant Physiology.* 139, 403-12.

227. Thomas, H. 1997. Chlorophyll: a symptom and a regulator of plastid development. *New Phytol.* 136 : 163-181.

228. Thomas, H., Stoddart, J.L. and Potter, J.F. 1980. Temperature responses of membrane associated activities from spring and winter oats. *Plant, Cell and Environ.* 3 : 271-277.

229. Thompson JE, Paliyath G. Brown JH, Duxbury CL. 1987. The involvement of activated oxygen in membrane deterioration during senescence. In: Thomson WW. Nothnagel EA. Huffaker RC. eds. Plant senescence: its biochemistry and physiology Rockville, MD: The American Society of Plant Physiologists, 146-55.

230. Titus, J.S. 1989. Nitrogen recycling in the apple (*Malus domestica* Borkh.). *Ann. Sci. Fores.* 46 (Suppl.) : 654-659.

231. Trebitsh T., Goldschmidt E.E., Riov J. 1993. Ethylene induces de novo synthesis of a chlorophyll degrading enzyme in *Citrus* fruit peel. *Proc. Natl Acad Sci.* USA 90: 9441-9445.

232. Tucker D. J. 1981. Phytochrome regulation of leaf senescence in cucumber and tomato. *Plant Sci Lett.* 23: 103-108.

233. Valpuesta C, Lange NE, Guerrero C, Reid MS. 1995. Regulation of a cysteine protease accompanies the ethylene-insensitive senescence of day lily (Hemerocallis) flowers Plant Molecular Biology 28, 575-82.

234. Valtersson, C., van Duijn, G., Verkleij, A., Chojnacki, T., de Kruijf, B. and Dallner, G. 1985. The influence of dolichol, dolichol esters and dolichyl phosphate on phospholipid polymorphism and fluidity in model membrane. J. Bio. Chem. 260 : 2742-2751.

235. van Staden J., Cook E.L., Nooden L.D. 1988. Cytokinins and senescence. In: Nooden LD, Leopold AC, eds. *Senescence and Aging in Plants.* San Diego: Academic Press, 281-328.

236. Vein, H. and van de Geijn, S.C. 1978. Mobility and ionic form of silver as related to longevity of cut carnations. *Planta.* 140 : 93-96.

237. Vicentini F., Hortensteiner S., Schellenberg M., Thomas H., Matile P. 1995a. Chlorophyll breakdown in senescent leaves: identification c of the lesion in a stay-green genotype of *Festuca pratensis. New Phytol.* 129: 247-252.

238. Vicentini F., Hortensteiner S. Schellenberg M., Thomas H., Matile P. 1995. Chlorophyll breakdown in senescent leaves: identification of the biochemical lesion in a stay-green genotype of *Festuca pratensis* Huds. *New Physiologist* 129, 247-52.

239. Vicentini F., Iten F., Matile P., 1995b. Development of an assay for Mg-dechelatase of oilseed rape cotyledons, using chlorophyllin as the substrate. *Physiol Plant.* 94: 57-63.

240. Vicentini F., Matile P. 1993. Gerontosomes, a multi-functional type of peroxisome in senescent leaves. *Journal of Plant Physiology.* 142:50-6.

241. Vierstra R.D. 1993. Protein degradation in plants. *Annual Review of Plant Physiology and Molecular Biology.* 44: 385-410.

242. Wagner E. 1976. The nature of photoperiodic time measurement: energy transduction and phytochrome action in seedlings of *Chenopodium rubrum.* In *Light and Plant Development* (Edited by H. Smith) pp. 419-443. Butterworths, London.

243. Wang, J.W. and Kao, C.H. 1999. Endogenous abscisic acid levels are not linked to methyl jasmonate promoted senescence of detached rice leaves. Plant Growth Regul. 28 : 17-20.

244. Wanner L. Keller F, Matile P. 1991. Metabolism of radiolabelled galactolipids in senescent barley leaves. *Plant Science.* 78, 199-296.

245. Wardle K. and Short K. C. 1983. Stomatal responses and the senescence of leaves. *Ann. Bot.* 52: 411-412.

246. Wareing, P. F. and Thompson A. G. 1976. Rapid effects of red light on hormone levels. In *Light and Plant Development* (Edited by H. Smith) pp. 285-294. Butterworths, London.

247. Watanabe H., Abe K., Emori Y., Hosoyama H., Arai S. 1991. Molecular cloning and gibberellin-induced expression of multiple cysteine proteinases of rice seeds (oryzains). *Journal of Biologial Chemistry.* 266, 16897-902.

248. Watson D.J. 1952. The physiological basis for variation in yield. *Advances in Agronomy.* 4, 101-45.

249. Weidhase, R.A., Kramell, H.M., Lehmann, J., Liebisch, H.W., Lerbs, W. and Parthier, B. 1987. Methyl jasmonate induced changes in the polypeptide pattern of senescing barley leaf segments. *Plant. Sci.* 51 : 177-186.

250. Welburn, S.C., Dale, C., Ellis, D., Beecroft, R. and Pearson, T.W. 1996. Apoptosis in procyclic *Trypanosoma brucei rhodesiense* in vitro. *Cell Death Differ.* 2 : 229-236.

251. Whitehead, C. S., Fujino, D.W. and Reid, M.S. 1983. Identification of the ethylene precursor, ACC in pollen. *Sci. Hort.* 21 : 291-297.

252. Willekens H., Langebartels C., Tire C., Van Montagu M., Inze D., Van Camp W. 1994. Differential expression of catalase genes in *Nicotiana plumbaginifoila* (L). *Proc. Nat. Acad. Sci.* US A. 91, 10450-4.

253. Wingate V.P.M., Lawton M.A., Lamb C.J. 1988. Glutathione elicits a massive and selective induction of the plant defense genes. *Plant Physiol.* 87, 206-10.

254. Woltering, E.J. and Harren, F. 1989. Role of rostellum desiccation in emasculated – induced phenomenon in orchid flowers. *J. Exp. Bot.* 40 : 907-914.

255. Woltering, E.J. and van Doorn, W.G. 1988. Role of ethylene in senescence of petals: Morphological and taxonomical relationships. *J. Exp. Bot.* 208 : 1605-1616.

256. Woltering, E.J., Harren, F. and Boerrigter, A.M. 1988. Use of a laser-driven photoacoustic detection system for the measurement of ethylene production in *Cymbidium* flowers. *Plant Physiol.* 88:506-510.

257. Woodson, W.R., Lawton, K.A. and Goldsbough, P.B. 1889. Ethylene-regulated gene expression during carnation petal senescence. Acta Hort. 261: 137-144.

258. Woolhouse H. W. and Jenkins G. I., 1983. Physiological response, metabolic change and regulation during leaf senescence. In *Growth and Functioning of Leaves* (Edited by J.E. Dale and F.L. Milthorpe) pp. 449-487. Cambridge University Press.

259. Woolhouse, H.W. 1984. Senescence in plant cells. In: Davis, I. and Sigee, D.C. eds. *Cell Ageing and Cell Death*. Cambridge: Cambridge University Press, 123-153.

260. Xing, T., Higgine, V.J. and Blumwald, E. 1997. Race-specific elicitos of Cladosporium fulvum peromote translocation of cytosolic components of NADPH oxidase to the plasmamembrane of tomato cells. *Plant Cell*. 9 : 249-259.

261. Yang HS, Huang HC, Collmer A. 1993. *Pseudomonas syringae* pv. *syringae* Harpin: a protein that is secreted via the Hrp pathway and elicits the hypersensitive response in plant. *Cell*. 73: 1255-66.

262. Zarembinski T.I., Theologis A. 1994. Ethylene biosynthesis and action: a case of conservation. *Plant Molecular Biology*. 26: 1579-097.

263. Ziegler R., Blaheta A., Guha N., Schonegge B., 1988. Enzymic formation of pheophoribide and pyropheophorbide during chlorophyll degradation in a mutant of *Chlorella fusca* Shihira et Kraus. *J. Plant Physiol.* 132: 327-333.

SUBJECT INDEX